The Alcohol

Fuel Handbook

By

Lynn Ellen Doxon

ISBN 978-0-7414-0646-0

Published by:

Infinity Publishing.com

Info@buybooksontheweb.com
www.buybooksontheweb.com
Toll-free (877) BUY BOOK
Local Phone (610) 520-2500
Fax (610) 519-0261

Printed in the United States of America

Published February-2013

TABLE OF CONTENTS

CHAPTER 1

INTRODUCTION

The Tallgrass Research Center was established in 1979 to investigate renewable energy sources and to encourage the small scale use of these resources. At that time national policy supported the investigation and use of renewable energy sources. In 1983, following the death of Robert Brautigam, cofounder and chief researcher as well as a change in the Federal administration that made renewable energy less of a priority nationally, the Tallgrass Research Center closed. This book went out of print at that time. It is being reprinted now because the author feels that the current world situation makes fuel alcohol more viable economically and more responsible socially than it has been at any previous time.

This book provides basic instruction on the production and use of fuel alcohol on the farm. It sets forth a simple system that can be built in a few hours. With the information in this book you can see for yourself whether it is worth the effort to produce your own fuel. This book does not intend to dictate any specific system. We show here what is probably the cheapest way to "learn by doing" at home. As you see, smell, touch and use alcohol you have actually produced you can adjust the procedures or design the system that best meets your needs.

This system is designed for the farmer with grain and livestock, although some other sources of raw material can be used that do not produce the high protein feed for animals that is a byproduct of grain alcohol production. With the information in this book we hope to guide you in building a simple system to produce small amounts of alcohol, give you some idea of what can go wrong and give you some basis on which to judge the feasibility of fuel alcohol production for yourself.

ALCOHOL HISTORY

We will teach you to make ethanol, backwoods moonshine. Man has been making it since before recorded history. Many centuries ago a cave woman probably left a jar of grape juice sitting in the sun by the entrance to the cave while she went to rescue the baby from a sabertooth tiger. When the man of the cave came home many hours later from a long day at the mastodon race he found the jar and drank the juice. It tasted a little funny but he felt pretty good when he had finished the jar. He asked his wife to prepare more of that the next day. His wife went out, gathered more grapes, made more juice and soon had more of the new drink. They shared it with the neighbors and before long the whole neighborhood was sitting down to a glass of wine with dinner. Those early winemakers probably never dreamed it was a living thing that was making their new drink. They did figure out, though, how to make other varieties of the drink and how to make vinegar or keep it from becoming vinegar. It was not until Louis Pasteur that anyone figured out what was actually happening. A microscopic fungus called yeast grows in sugary liquid and produces alcohol and carbon dioxide.

Just as alcohol has been produced since civilization began governments have been making laws concerning it since government began. The laws and taxes current at the time of publication for residents of the United States are discussed later in this book. New regulations are being made every day and the agents of the Bureau of Alcohol, Tobacco and Firearms are very cooperative about assisting with fuel permits.

Our bodies quickly make energy out of alcohol. Our engines can do the same thing. Of course, our engines will not go blind or die because of small amounts of contaminants in the batch, so we don't need to be quite as careful as the big distilleries that make drinking alcohol.

The first fuel used in the internal combustion engine was alcohol. Shortly after the internal combustion engine was invented petroleum distillation was discovered. At that time gasoline was much cheaper to produce than alcohol, there was little concern over air pollution and oil supplies were thought to be inexhaustible. Only a few foresighted people realized the disadvantage of using a fuel that had to be searched for and mined from underground. Henry Ford was one of these. He was a farm boy and knew that the farm had traditionally produced energy in the form of animals and animal feed as well as producing food. He realized that removing energy production from the farm would cut in half the economic base of farming and fought long and hard for the use of alcohol as fuel.

OVERVIEW OF ALCOHOL PRODUCTION

Making alcohol is not far removed from chores farmers are used to. It is growing a yeast crop for the alcohol it produces. Grain is ground to make the starches more available. Enzymes are then added to break the starch down to sugars. These are the same sorts of enzymes that are found in saliva. The sugar is then fed to yeast that digests the sugar and water to produce alcohol and carbon dioxide (along with more little yeast organisms.) The yeast finally starves to death or kills itself off by overpopulation and too much alcohol. We then remove the liquid, which is alcohol and water, and distill it. The solids - the protein that was in the grain and the dead yeast organisms - are fed to animals as a protein supplement.

Substrates

The substrate is the material from which the alcohol is made. If you were just starting to farm, without any land or equipment, you would go out and look for land that will grow the crops you are interested in and that you can afford. Rich, black bottom land will grow more than rocky, yellow hillsides. Carbohydrates are what make an alcohol crop. Sugar and starch are carbohydrates. Crops with more carbohydrates will produce more alcohol per pound. Table I-1 gives the amount of alcohol that can be produced from several different crops.

If you are buying the substrate calculate the cost of the alcohol by dividing the cost per unit by the number of gallons that unit will produce. For example, lets say you want to produce alcohol from pure cane sugar and you can get that sugar at $8.65 per 100 pounds.   You can make 6.92 gallons of alcohol from that sugar so the cost would be $1.25 per gallon of alcohol. If you were buying wheat at $4.50 a bushel and could make 2.56 gallons of alcohol from that wheat the cost of the gallon of alcohol would be $1.75. This does not include a cost for your time or investment return.

If you are growing the crop yourself the more carbohydrates per acre the more alcohol per acre. If a crop will produce many gallons of alcohol per bushel but will only produce a few bushels per acre it might be better to choose another substrate. To figure the amount of alcohol per acre multiply the average production per acre by the amount of alcohol that crop can produce. (Make sure the units are the same.) In order to figure the cost of the substrate for each gallon of alcohol divide the cost of production per acre by the number of gallons that can be produced from the substrate grown on that acre. For example, if you can grow 40 bushels per acre of dryland wheat, which will produce 2.56 gallons of alcohol per bushel, the yield will be 102.4 gallons per acre. At a production cost of $55 per acre the substrate will cost $.54 per gallon.

There are several things to consider when deciding what substrate to use. In addition to expense, you should consider how dependable the crop is in your area, whether the equipment is available to plant, care for and harvest the crop, whether you can store it until you are ready to use it and whether you have the equipment to prepare it. Will you use the culls from your potato or fruit crops? Will you plant what would once have been your set aside acres into grain? Will you use different crops at different times of the year? Each operation is different and you must decide for yourself what is best.

There is a residue left over after the alcohol is made that is two to four times as rich in protein as the material going in. Certain other nutrients are concentrated also. With some substrates this is a high quality, high protein animal food. With others it is not usable. Could you formulate a supplement for your animals that would provide an amino acid balance? More on that in the chapter on byproducts.

Just as there is a need for water to make the nutrients in the soil available for plants, water must be available to dissolve the carbohydrates in the substrate you use. The concentration of carbohydrates in the liquid should be between 10% and 25%. The more water you add the more complete fermentation takes place. The less water the more concentrated the alcohol in the brew. The yeast will die when the concentration of the alcohol gets to be around 12%, so do not increase the sugar concentration above 25%. If there is less water there are some carbohydrates left in the feed product and the protein is not so concentrated. A simple way to figure how much to add is to multiply the percentage fermentable in Table I-1 by ten and subtract 100. Add that many pounds of water to 100 pounds of substrate. Remember, "a pints a pound the world around." (That's close enough, anyway.) As you get experience with distillation you might want to adjust that figure somewhat. We find 30 gallons of water per bushel of milo a convenient amount.

Before planting a crop you would plow or disc to prepare a seedbed. The seedbed must also be prepared in alcohol production. The pieces of the substrate must be small enough that the enzymes and yeast can get to the carbohydrates. Grain should be ground to the consistency of coarse cornmeal. Other substrates should be ground, mashed or shredded as appropriate to the substrate. Save all the juices. Sugar is water soluble and a lot of it could run off with the juice of some crops. One half of the water should be added to substrates containing starch and they should be heated to soften the starch. If you bring it to a boil for a short period of time you will also kill unwanted bacteria and other microscopic weeds that would disrupt the production of alcohol.

Table I-1. COMMERCIAL AVERAGE YIELD OF 200 PROOF ALCOHOL

Material	Unit	Lbs./Unit	% Fermentable	Gal./Unit
Wheat	Bushel	60	58.6	2.56
Corn or Milo	Bushel	56	57.8	2.34
Rye	Bushel	56	54.0	2.19
Buckwheat	Bushel	48	57.2	1.99
Barley	Bushel	48	54.3	1.89
Oats	Bushel	32	43.6	1.01
Sugar beets	Ton	2000	16.0	22.00
Sugar cane	Ton	2000	11.0	15.18
Sweet potatoes	Bushel	55	23.3	.93
Potatoes	Bushel	60	15.6	.68
Jerusalem Artichokes	Bushel	60	15.2	.59
Pure sugar	Bag	100	100.0	6.92
Corn sugar	Bag	100	100.0	6.97

Enzymes

Yeast makes alcohol from sugar. Starches are long chains of sugar and cellulose is a mass of starches cemented together. Enzymes can break down both starch and cellulose into sugar. The enzymes that break down starch are readily available commercially. Those that break down cellulose are less available.

Enzymes are almost magical compounds that make biological processes different from all other processes. They make things happen that would not ordinarily happen. Enzymes do not get used up in the reaction they make happen, although there are many things that will inactivate them. A given enzyme will do only one thing. It is like a key that will fit only one lock. They act best at a certain temperature and pH. pH is a measure of the acidity or alkalinity of a solution. It is determined by measuring the concentration of hydrogen or hydroxide (OH-) ions in a solution. More hydroxide ions mean higher pH or more alkaline. More hydrogen ions mean lower pH or more acidic. Distilled water is neutral - has an equal number of hydrogen and hydroxide ions. Its pH is 7.0.

Enzymes found in saliva, sprouted grain and certain bacteria break the bonds that hold sugar together in starch chains. The enzymes are called amylases. There are two different kinds of bonds between sugars in starch chains. To break these bonds it takes two different amylases. They are glucoamylase and alpha-amylase. The difference between the two is like the difference between right and left-hand scissors. They approach from different directions and therefore are able to cut slightly different bonds. In this book we discuss Diazyme, a glucoamylase and Taka-therm, an alpha-amylase. Both are from Miles Laboratory. Similar enzymes are available from other companies.

There are enzymes called cellulases that break the sugars in cellulose apart. These are produced by bacteria in the first stomach of ruminants and are very expensive to buy at this time. A method called acid hydrolysis can also be used to break down the cellulose, although there is currently not a simple, safe version of this process that you can do at home.

Taka-therm is most active at a pH range of 5.5 to 7.0, or in slightly acid to neutral conditions. This can be measured with pH paper available at most drug stores. It will retain its ability to act in a pH range of 5.0 to 11.0. It works best at temperatures below 194° F. If calcium ions are present in the water it will act at higher temperatures, although higher temperatures for long periods of time still tend to inactivate the enzyme. It should be stored at low temperatures (40°F).

Diazyme works best at a pH of 3.8 to 4.2 and at 140° F. It will work in a pH range of 3.5 to 5.0. Temperatures above 176° F. will inactivate this enzyme.

When grain has been ground, mixed with cool water and heated you should add one ounce of Taka-therm per bushel. The grain should be mixed with cool water at first to prevent balling but once it is moistened you may add water from some other step in the distillation process. This will save on energy for heating. The Taka-therm can be added any time after the heating process has begun. The heat causes the starch to leave the grain and turn the mash to gel. The Taka-therm liquefies this gel. As you heat, agitate the mixture so it will not scorch the grain and the heat will transfer faster.

After the Taka-therm has had at least an hour to work, mix cooler water with the mash until it is about 140° F. If you add all the water at this time and it is still not cool enough just let it set until it is. Again, test the pH and adjust it to between 3.5 and 5.0. To lower the pH add any acid. Battery acid or muriatic acid is probably the most available. To raise pH add lime. It is unlikely that you will need to raise the pH before adding the Diazyme if you are careful as you add the acid. Add ½ to 2 ounces of Diazyme per bushel and let it work at least half an hour. If you have not added all the water do that when the Diazyme has had a chance to work. Let the mixture cool to 90°F.

Fermentation

Brewer's yeasts are related to mushrooms, although they are smaller, and are more closely related to the organisms that give us penicillin. They are in the air all around us. Any natural fermentation is caused by one variety of yeast or another. There are a variety of other living things - fungi, bacteria, viruses - floating around in the air, too. Since it was discovered that yeast produces alcohol producers of alcohol have worked on varieties and systems to make sure alcohol is produced and the byproducts of other systems are not. We can benefit from the many years of research that have been done. High temperatures kill any living thing. Therefore we have killed any natural yeast and other living organisms in the mash when we cooked it. We must add yeast if we are to get any fermentation action.

Yeast companies have developed strains of yeast that ferment sugars swiftly and efficiently. Therefore we can get the desired results more quickly than if we try to capture and breed our own. The product called Brewer's yeast has materials added to retard the growth of weeds in our brew as it makes alcohol.

Yeast will die at temperatures above 115°F. Be sure you do not put the yeast in while the mixture is too hot. They are most active near 90°F. They are living things and produce heat by the action of living. The fermentation vat will stay warm if it is insulated and inside a building but care must be taken to prevent it from overheating on hot summer days.

Dissolve the yeast in a small amount of water between 90° F. and 115° F. and mix it with your mash. You should use two ounces of yeast per bushel of grain. Once you start a batch with a fast acting variety you can keep on producing your own seed until it gets contaminated with weeds.

How long that will be depends on how clean you keep your seed jar and equipment. Just take a jar, fill it with some of the grain and water from the fermentation vat and let it sit, stirring frequently. The more oxygen it gets the faster the yeast reproduces.

Yeast has two biological pathways. If the yeast has oxygen it will reproduce and make large quantities of carbon dioxide. If it does not have oxygen it will produce alcohol and smaller quantities of carbon dioxide. Therefore you should exclude oxygen from the fermentation vat. You must allow carbon dioxide to get out. We achieve this by covering our fermentation vats with plastic, held in place with a rubber band cut from an innertube.

The yeast is working as long as there is grain on top of the liquid and small bubbles coming up. When the grain falls to the bottom, after two to four days, the batch is ready to distill.

If there is no grain at the top and large bubbles are coming up from the bottom, you probably have a vinegar producing organism in your brew. This is the most common weed found in alcohol crops. The vinegar producing organisms live on alcohol - the longer they are there the less alcohol you have in your brew. When you notice them, either distill or discard the batch immediately and clean out the fermentation vat with boiling water. There will not be as much alcohol as usual in the brew but if the batch is one or two days old and the vinegar hasn't been there long it would be worth distilling.

Distillation

Distillation is the separation of two compounds using heat. By heating the liquid to the point where it boils then cooling it slowly the water recondenses first and we can then condense and collect the alcohol.

USES OF ALCOHOL

Once you have the alcohol, certain modifications need to be made in engines designed to burn gasoline in order to accommodate alcohol fuel. These changes are due to the facts that alcohol is thicker than gasoline, burns more completely and burns cooler. The modifications are described in the chapter on fuel use. There are engines designed to run exclusively on alcohol. These are, or have been made by American-based companies but are not available for sale within the United States.

Alcohol can be used in any heater or furnace that burns fuel oil without any modifications. The furnace can then be vented directly into the house, allowing you to use all the heat produced by the fuel rather than sending over half of it up the chimney. You will need to provide an oxygen intake from outside if your furnace does not have one. You can also burn alcohol as low as 130 proof. The burning alcohol will evaporate the water and the air coming out of the stove or furnace will be warm and moist, a very pleasant humidifier.

Figure I-1. Weight gains by cattle fed distillers grain.

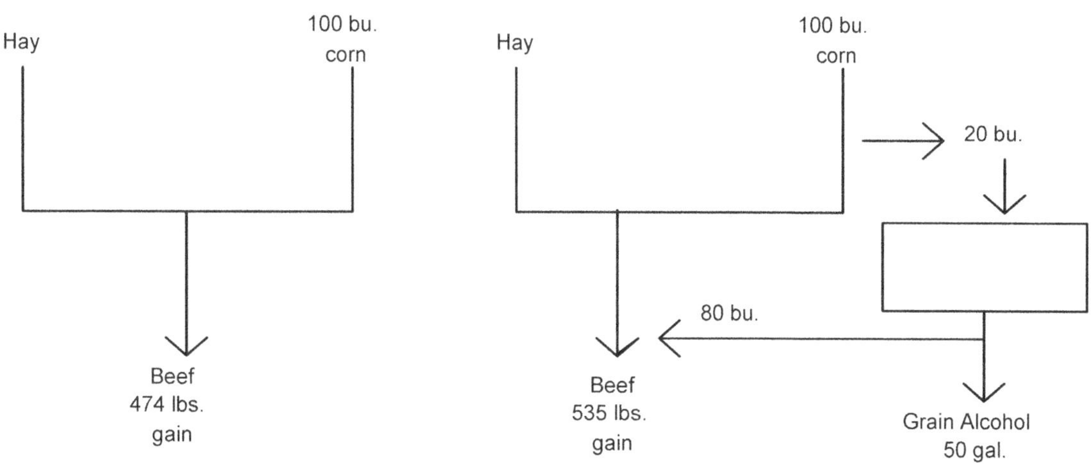

12.
9% increase in gain plus 50 gallons alcohol. Based on research reported by W.P. Garrigus, University of Kentucky, at the Tenth Distillers Feed Conference, Cincinnati, Ohio, March 3, 1955.

USES OF DISTILLERS GRAIN

In addition to producing alcohol, you will be producing a high protein feed product and excess heat. The feed product is almost equal to soybean meal as a protein supplement for animals. It is not a complete ration and must be mixed with other feedstuffs. The yeast in the product and the fact that it bypasses the rumen increases the efficiency of utilization by ruminants. Figure I-1 shows the results of 1955 research on this subject. Several universities have worked on further research in more recent years. Depending on what is mixed with the feed product it can be fed to cattle, sheep, goats, horses, swine, chickens, dogs or fish with excellent results. It can even be added to bread, cereal or baked goods for people.

USES OF HEAT BYPRODUCTS

Despite some concern about heat loss, winter could very well be the best time for making fuel alcohol. Farmers usually have more time in the winter. It is also easier to control the fermentation temperature at this time of year. In the summer we may have to cool the fermentation tank to keep it from overheating and killing the yeast. In the winter, simply retaining the heat by insulating the tank and putting it inside a tight building we could maintain the proper temperature. Every time we cook a batch there is considerable heat loss from the cooker that could be used to maintain the proper temperature in the room. Every time we distill there is heat loss as well as hot water produced. In the winter this heat can be utilized in the alcohol production area, adjacent building or your home. In the summer it would be wasted and could overheat the fermentation vats.

How well this heat is utilized will depend on how you set up your system. It could be that you could completely write off the cost of the fuel necessary for alcohol production because the heat it produces in used in other areas.

PRODUCTION POTENTIAL

Table I-2 shows the amount of alcohol you could get from various substrates if all the carbohydrates were converted to alcohol and carbon dioxide. This table is for comparison purposes only. Do not expect to get these yields from 100 pounds of the substrate in actual practice. For example, 80% of the sugars can generally be extracted from grapes, while only 75% can be extracted from apples. In actual practice a good yield is anywhere between 70% and 90% of these figures.

Sweet sorghum is not included in this table because of a very wide variation in the sugar content of different varieties. Different varieties of the products listed also have differing sugar contents but the variation is not quite as large as the variation in sweet sorghum.

We have not fermented a number of these products and therefore cannot give you specific suggestions on most of them. In general fruit had the greatest amount of sugar when it is _very_ ripe but not rotten. The same would be true of vegetables. It is generally best to remove the juice from the pulp and ferment only the juice for the sake of your equipment, but this is not essential to good alcohol production. The juice should be removed from stems and weeds in products like sweet sorghum and sugar cane so that the byproduct is easier to dispose of and to avoid tannic acid contamination.

The more moisture in dried items (grains, beans, etc.) the easier they are to cook and prepare for fermentation. However, if they are to be stored be sure they are dry enough to survive storage without heating or molding, which could reduce your yield. If you do have moldy or spoiled grain or beans you can still get some alcohol out of them, so do not throw them away.

To compare the potential of other products, find out the percent carbohydrate in the product and divide by 13.7. This will tell you the maximum number of gallons of alcohol per 100 pounds of the substrate.

Table I-2: WATER, CARBOHYDRATE AND ALCOHOL CONTENT OF SELECTED FARM PRODUCTS.

	% Water	% Carbohydrates	Alcohol
Apples, raw	84.4	14.5	1.08
Apricots, raw	85.3	12.8	.95
Artichokes, French	85.5	10.6	.79
Artichokes, Jerusalem	79.8	16.7	1.24
Asparagus, raw	91.7	5.0	.37
Beans, lima, dry	10.3	64.0	4.67
Beans, white	10.9	61.3	4.47

	% Water	% Carbohydrates	Alcohol
Beans, red	10.4	61.9	4.52
Beans, pinto	8.3	63.7	4.65
Beets, red	87.3	9.9	.72
Beet greens	90.9	4.6	.33
Blackberries	84.5	12.9	.94
Blueberries	83.2	15.3	1.11
Boysenberries	86.8	11.4	.83
Broccoli	89.1	5.9	.43
Brussels Sprouts	85.2	8.3	.60
Buckwheat	11.0	72.9	5.32
Cabbage	92.4	5.4	.38
Carrots	8.2	9.7	.70
Cauliflower	91.0	5.2	.37
Celery	94.1	3.9	.28
Cherries, sour	83.7	14.3	1.04
Cherries, sweet	80.4	17.4	1.27
Collards	85.3	7.5	.54
Corn, field	13.8	72.2	5.27
Corn, sweet	72.7	22.1	1.61
Cowpeas	10.5	61.7	4.94
Cowpeas, undried	66.8	21.8	1.59
Crabapples	81.1	17.8	1.29
Cranberries	87.9	10.8	.78
Cucumbers	95.1	3.4	.25
Dandelion greens	85.6	9.2	.67
Dates	22.5	72.9	5.32
Dock, sheep sorrel	90.9	5.6	.40
Figs	77.5	20.3	1.48
Garlic cloves	61.3	30.8	2.25
Grapefruit pulp	88.4	10.6	.77
Grapes. American	81.6	15.7	1.14
Lambs-quarters	84.3	7.3	.53
Lemons, whole	87.4	4.9	.35
Lentils	11.1	60.1	4.38
Milk, cow	87.4	4.9	.35
Milk, goat	87.5	4.6	.33
Millet	11.8	72.9	5.32
Muskmelons	91.2	7.5	.54
Mustard greens	89.5	5.6	.40
Okra	88.9	7.6	.55
Onions, dry	89.1	8.7	.63
Oranges	86.0	12.2	.89
	% Water	% Carbohydrates	Alcohol

Parsnips	79.1	17.5	1.27
Peaches	89.1	9.7	.70
Peanuts	5.6	18.6	1.35
Pears	83.2	15.3	1.12
Peas, edible pod	83.3	12.0	.87
Peas, split	9.3	62.7	4.57
Peppers, hot chile	74.3	18.1	1.32
Peppers, sweet	93.4	4.8	.35
Persimmons	78.6	19.7	1.43
Plums, Damson	81.1	17.8	1.20
Poke shoots	91.6	3.1	.23
Popcorn	9.8	72.1	5.27
Potatoes, raw	79.8	17.1	1.24
Pumpkin	91.6	6.5	.49
Quince	83.8	15.3	1.12
Radishes	94.5	3.6	.26
Raspberries	84.2	13.6	.99
Rhubarb	94.8	3.7	.27
Rice, brown	12.0	77.4	5.64
Rice, white	12.0	80.4	5.86
Rutabagas	87.0	11.0	.80
Rye	11.0	73.4	5.35
Salsify	77.6	18.0	1.31
Soybeans, dry	10.0	33.5	2.44
Spinach	90.7	4.3	.31
Squash, summer	94.0	4.2	.30
Squash, winter	85.1	12.4	.90
Strawberries	89.9	8.4	.61
Sweet potatoes	70.6	26.3	1.91
Tomatoes	93.5	4.7	.34
Turnips	91.5	6.6	.48
Turnip greens	90.3	5.0	.36
Watermelon	92.6	6.4	.47
Wheat, HRS	13.0	69.1	5.04
Wheat, HRW	12.5	71.7	5.23
Wheat, SRW	14.0	72.1	5.26
Wheat, white	11.5	75.4	5.50
Wheat, durum	13.0	70.1	5.11
Whey	93.1	5.1	.37
Yams	73.5	23.2	1.69

It is possible for you to produce your own enzymes. However, this is not something we recommend except through absolute necessity. The enzymes we buy are extracted from *Bacillus lichenformus* and *Aspergillus niger*. To produce your own enzymes you could grow cultures of these and extract the enzymes or grow malt. According to the analysis of the biochemical activity of the enzymes produced in each case, and the activity of the yeast, a higher alcohol yield is possible using the bacterial enzyme than using malt. However the production of the enzymes in less than laboratory conditions is difficult enough to offset any benefit.

To grow the bacterial culture you need a substrate, just as in growing yeast. *Bacillus licheniformus* is a bacteria and *Aspergellus niger* is a *Fungi Imperfecti*, or almost a fungus. Each will produce high quantities of starch to sugar enzymes if grown on a high starch substrate. They excrete the enzymes, which make glucose from the starch, then use the glucose as food. Therefore the enzyme must be extracted from the living organism rather than putting the organism directly in the fermenting brew, or the organism will use the sugars for food before the yeast can get to it. Removing the enzyme requires a centrifuge, so it is expensive.

Even if you have a centrifuge, there is the problem of contamination to consider. *Aspergillus niger* is of the same genus as many organisms that cause spoilage. A substrate that will support the growth of one species will generally support the growth of another species of the same genus, so there could be a marked increase in the amount of spoilage around the farmstead if these cultures are allowed to become contaminated. The major contamination problem comes with *Bacillus licheniformis*. It is of the same genus as the bacteria that causes anthrax. So the backyard microbiologist could be growing a deadly *Bacillus anthrax* culture when he thinks he is growing a harmless *Bacillus licheniformus* culture.

If the enzymes are unavailable to you, making malt is probably the safest, easiest way to go, but even that takes time and attention. Malt is simply sprouting grain. Barley works best for malt (produces more enzyme) although wheat is good, too and any grain will produce some enzyme. Malt should consist of 10 to 20% of the grain in the mash.

The first step in making malt is to soak the grain. Pour fresh water over the grain until the water is six inches above the surface of the grain. If this is done during the summer or in a warm room the water should be changed every four to six hours so the grain does not start to spoil or ferment. It has soaked long enough when the grain can be crushed and leave no hard starch.

Next it should be germinated. Germination takes place with the fewest problems if it is in a room about 55° F. The soaked grain should be piled up to two feet deep. Within 12 to 24 hours the grain will begin to produce heat. This means it is beginning to grow. When the center of the pile gets warm and wet it should be turned. This turning should take place every six to eight hours after the heating has started. As the grain sprouts it will have to be turned more and more often to keep the center of the pile around 65° F. The piles should also be spread out to become thinner and thinner until they are only a few inches deep.

Small rootlets will form first. The sprouting should be stopped when these rootlets are about 2/3 the length of the grain. If leaflets are allowed to form they will use up a lot of starch that could otherwise be turned into alcohol.

It is very important that once the enzymes are formed the sprouting process is stopped quickly. Otherwise you will have a putrid mess. To stop the process it should be dried with heat. Start the drying process at around 95°F then gradually warm it up to 160°F. When it is nearly dry and feels dry to the touch you may raise the temperature higher to dry the malt completely. It can get to 212° at this point because the enzyme can only be inactivated at boiling temperatures if there is water present.

If you were making beer for drinking you would then separate the dried rootlets, which are brittle and easily winnowed out. Since you don't care how the alcohol tastes you can go ahead and grind everything you have dried and mix it with your starchy substrate. You now have beta-amylase, which will turn your starch to maltose, which the yeast can turn to glucose for use in making alcohol. Do not heat the malted mixture above 170°F after water is added or you will inactivate the enzyme.

CHAPTER 2

MONEY AND ENERGY

Probably the most common question we have been asked is "How much does it cost to produce a gallon of alcohol?" In this age, almost everything we do hinges on money. Another common question is "Can you produce more energy than you use?" Many so called "experts" have made these the central issues of alcohol production. There are no experts in the area of on-the-farm fuel alcohol production. There are alcohol industry experts who know all about making alcohol in factories with large stills. We can learn a lot from these people but we cannot copy their methods exactly or take their figures and transfer them exactly. There are some old moonshiners who have made many hundreds of gallons of relatively low proof alcohol in the backwoods. We can learn a great deal from them, too, but there are some differences between what they did and what we are trying to do. Finally there are college professors and independent researchers who have taken figures from industry and, without ever having made alcohol, have scaled them down to a smaller operation. It comes out looking very tidy but is no more accurate than Swift's account of Gulliver travels in Lilliputia.

In this chapter we will deal with the issues of cost and energy. We will simply do a comparison of monetary input and energy flow. We haven't given exact figures here because those will change by the time you read this. Your particular situation will also probably have some different factors to be considered that make change the economics for you.

COST

First, compare the two circles in Figure II-1. These compare the energy flow on two different farms. On each a feed crop is harvested. On Farm A, part of this feed crop is sold for processing and part goes into storage. The farmer then buys back the processed feed crop in the form of a protein supplement,. (Naturally at a higher price than he got for the original feed product.) He blends it into a balanced ration which he feeds to his livestock. Then he sells the livestock and hopes for a profit.

On Farm B the farmer harvests his feed crop and takes all of it to storage. His storage costs may be as much as 20% higher than those of Farmer A. He then processes a portion of the feed crop himself to produce alcohol. If he pays himself or someone else for the labor involved in this process it will cost about 50% more than the difference between the feed Farmer A sold and the protein supplement he bought. Animals gain faster on distillers grain. There would be about a 4% increase in gain, overall, in the animals it would take to balance the system. So, halfway through the circle Farmer B has spent 20% more on storage and 50% more on protein supplements and only made 4% more gross profit on his cattle. Let's finish the circle.

Farmer A has a manure disposal "problem." Many farmers today are returning to putting manure on the fields, realizing how much they can save on fertilizer this way, but a surprising number of them still consider it a problem and have elaborate disposal systems. Farmer B has a methane digester. This costs about the same as the elaborate disposal system. All the manure Farmer B can recover goes into his digester and out of it he gets the fuel to provide the heat for the protein processing. Farmer A goes out in the spring and buys fertilizer. Farmer B spreads digester sludge on his fields. Combined with alfalfa rotation this saves 80% of his fertilizer costs.

We have now arrived at planting time for the feed crop. Farmer A buys his fuel, discs, harrows and plants. Farmer B already has free fuel (all the costs were absorbed in the protein processing.) So how does the balance sheet come out?

Figure II-1

Energy Flow Diagram

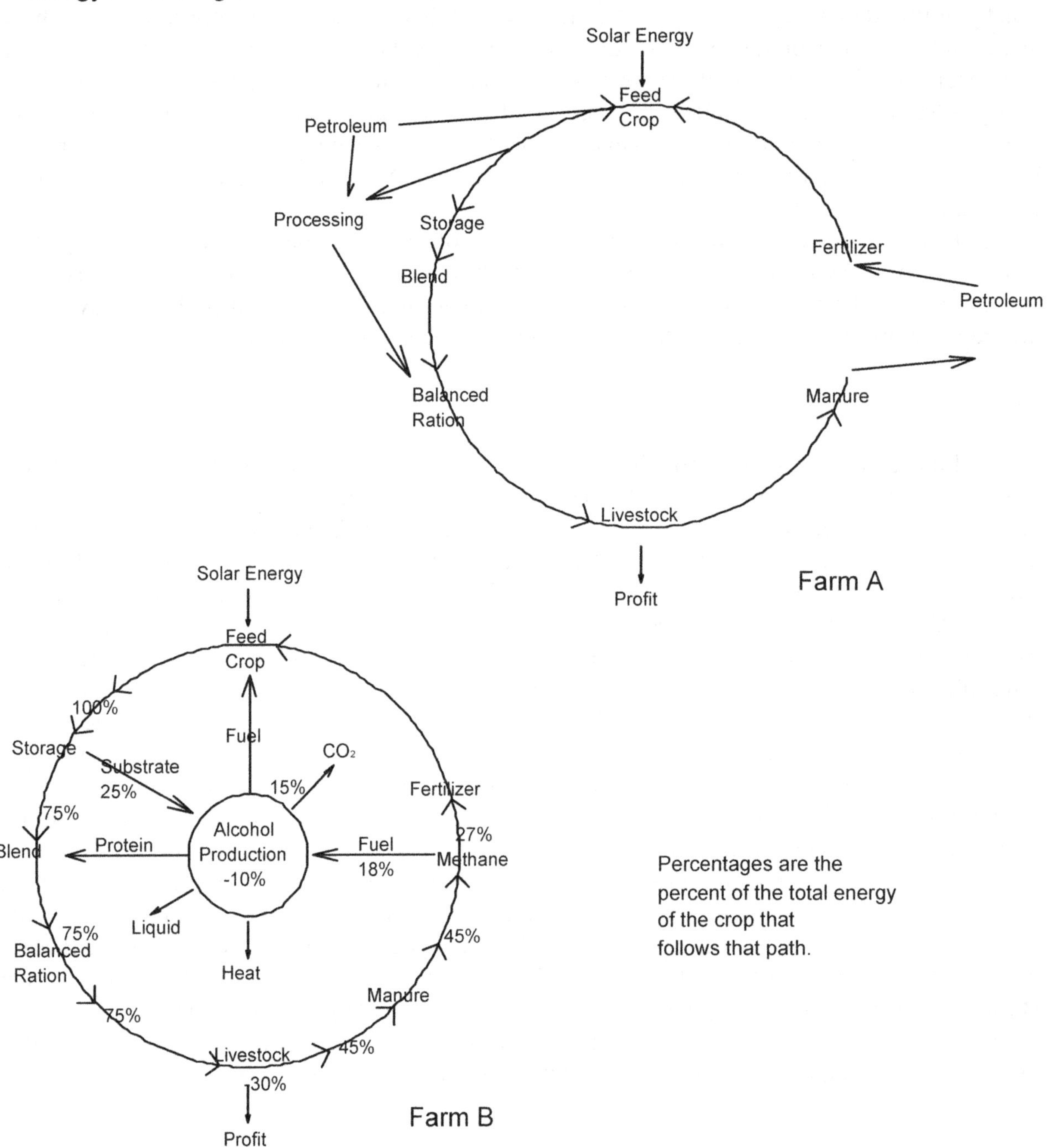

16

<table>
<tr><td>Farmer A</td><td>Farmer B</td></tr>
<tr><td>20% of storage costs</td><td>4% livestock profit</td></tr>
<tr><td>50% of protein costs</td><td>80% of fertilizer costs</td></tr>
<tr><td></td><td>100% fuel costs</td></tr>
</table>

Which farmer came out ahead? Farmer B also has more humus in his soil, adding ability to absorb water and ease of tilling. Preliminary research shows that his equipment will last 25 to 50% longer, his ability to farm does not depend on the whim of OPEC and there are few adverse effects on the environment.

Of course to come out that well the system must be balanced. If you don't have enough land, you may have to buy grain. If you don't have enough animals or can't collect their manure you will have to provide another fuel for the process. (This could be solar or wind.)

Table II-1 shows a cost comparison for different types of operations. You can see why commercial figures will not work on the farm. These costs are discussed below.

Table II-1. Cost comparison of different kind of operations.

Cost Description	On Farm	Feedlot	Distillery
Investment	X	X	X
Labor	X	X	X
Production Materials	X	X	X
Bonding	▽	X	X
Substrate	▽	▽	X
Fuel	▽	▽	X
Alcohol Transportation		X	X
Grain Transportation		▽	X
Commercial Costs		▽	X
Protein Drying			X
Protein Transportation			X

X = Cost involved in the process
▽ = Cost that may be involved in the process

Investment - To figure this cost on a per gallon basis divide the cost of the facility by the total number of gallons it is expected to produce in its lifetime. Because a feedlot would probably use a large, expensive still but would produce less than a commercial distillery the cost would likely be highest for the feedlot. The farmer can make do with much of his equipment gleaned from other farm operations. The time for the investment to be returned might be shortest for the distillery and second for the feedlot because they would be using the still continuously while the farmer would use it only when needed and the still would frequently sit idle. Commercial regulations require the use of additional equipment, so the initial cost is higher for any commercial producer.

Labor - This is a major factor for the distillery and the feedlot. Any continuous operation needs continuous monitoring. The larger the operation the more employees. On the farm alcohol production could be worked in among the morning and evening chores or done in the winter when there is less to do on the farm. Someone should be in the vicinity of the still when it is operation, but they could be fixing machinery in the shop where the still is working.

Production materials - This is one area where the distilleries come out ahead. They can get enzymes and yeast at bulk prices. This means that while they could get enough to make a gallon of alcohol for three cents you might have to pay as much as four and one half cents. Because this cost is so small it does not make farm production seem unattractive.

Bonding - Any operation that produces more than 10,000 proof gallons a year must post a bond with a bonding company. The amount of the bond is $2000 to $200,000 depending on production.

Substrate - This is the farmer's and feedlot's biggest advantage over the distillery. The farmer is using part of the crop he would feed to his animals anyway and is processing it in order to balance the ration and increase weight gain. He has only the production costs to consider, not the profit to the grower (when there is any) and the profit to anyone who handles it between the grower and the delivery point. All the costs are reduced by using the alcohol on the farm to produce the grain. Should he decide to do so, the farmer can assign all the production costs to the protein supplement he is making for his animals (because they are sold and bring cash into the farm) and completely wipe out his fuel costs.

Fuel - The farmer can use solar heat, methane , wind generated electricity or other products that would otherwise be wasted on the farm. A feedlot is in the ideal position to use methane. Because the cost of the fuel for making alcohol is one of the greatest costs you can see that the distilleries are at a disadvantage. Also, because of the design of most commercial distilleries and the fact that they dry the distillers grain, they use more fuel per gallon that the farmer or feedlot would.

THE ABOVE ARE THE ONLY COSTS THE FARMER WOULD NORMALLY CONSIDER.

Alcohol Transportation - The distillery and the feedlot are making more alcohol than could conveniently be used on the premises. The feedlot might deliver the extra alcohol to farmers in the area who would supply him with grain. There is some cost - or at least some use of fuel for this. A distillery might have a nationwide market and their cost for getting their products to that market would be considerably higher.

Grain Transportation - Here again the farmer would be carrying the grain the same distance, or even a shorter distance than he normally would. The feedlot would have to bring the grain at least from the neighboring farms, something they would have to do anyway to feed the cattle. Distilleries are generally in a town some distance from their grain supply so they would have the largest grain transportation costs.

Commercial costs - Besides the additional costs of a bond a commercial operation has to denature the product. Denaturing compounds are expensive, even when you use very small quantities. Commercial installations are much more concerned with quality control than the farmer because the customer must be satisfied. The cost of building, supplying and staffing a quality control laboratory can be astronomical. ATF strictly controls commercial establishments and they must have their own inspectors and record keepers. As you well know, red tape is one of the most expensive commodities available today. All of these factors add considerably to the cost of a gallon of commercially produced alcohol.

Protein drying - Only the distillery would be concerned with this. The protein is too valuable to waste, so distilleries dry it and sell it. They also have to transport it to market.

ENERGY

We all learned in high school that energy cannot be created or destroyed, only changed in form. We also learned that it takes the same number of BTUs to boil a gallon of water no matter how it is done. The exact amount of energy it takes to distill a gallon of alcohol will depend on the efficiency of your fermentation process and the design of your still. A still that heats water first will take more energy than one that heats the brew directly. A heat source that is inefficient will take more energy than one that is efficient. You will be able to get some idea of the efficiency of using different fuels in the chapter on heat. In general, the amount of energy it takes to distill a gallon of alcohol will be on the same order of magnitude for all stills. A still that will not produce three BTUs of energy for every BTU that is used is inefficient.

Go back to Figure II-1. We will examine the energy flow on these two farms. Both crops absorb solar energy and convert it to chemical energy. Fuel is then used to harvest the crop and take it to storage or the market. On Farm B 25% of the crop is processed into alcohol. The yeast uses 40% of the energy in that portion, or 10% of the total energy of the crop and the rest, 15% of the total - is given off as alcohol. Petroleum energy is also solar energy converted into fuel but it took millions of years and did not take place on the farm where the fuel is used.

On Farm A part of the feed crop is processed off the farm (again using petroleum fuel for processing and transportation) while on Farm B it is processed on the farm. The balanced ration that results from all this processing and blending is fed to livestock. They remove some of the energy (about 40% of the feed or 30% of the total). So we have 45% of the original energy of the plants in the manure. By putting this manure in a digester we can extract another 40% of that energy in the form of methane. Forty percent of the 45% or 18% of the total solar energy absorbed can go into the distilling process. It would take only 10% of the total energy, at most, to distill the alcohol produced above. The extra methane could be used to generate electricity or cook food in the house. The heat used in the distillation process is not used up, either, as you will see later in the book. It can be used to heat a house, shop, farrowing barns or anything else you like.

When the methane is made 27% of the original energy is returned to the field. Energy, at this point, is carbon. Having removed this carbon has little effect on the soil. The limiting factor in crop growth will be nitrogen, potassium, phosphorus or in some cases iron or zinc. The cattle probably removed 18% to 25% of the nitrogen for growth and survival and the methane process probably removed .005 to 1%. This means we end up returning 74% to 80% of the nitrogen to the field. In addition to applying the sludge Farmer B rotates the grain crop with alfalfa (the portion of his feed crop not processed into fuel.) Therefore Farmer B has saved almost all of his fertilizer costs. (Occasionally rock mineral applications are necessary.)

On Farm A many of these processes were done with petroleum energy. There are more BTUs in a gallon of gasoline than in a gallon of alcohol. The efficiency of their use will be discussed later. At this time I would like to comment on the energy investment in that petroleum fuel. It takes energy to get the oil out of the ground and to the refineries. It takes energy to refine the oil into the various products we use. The major difference between alcohol distillation and petroleum distillation is that petroleum distillation must reach a temperature of $720°F$. Alcohol distillation never gets above

212°F., the boiling point of water. Thus the distillation process for gasoline takes almost three times the energy that the distillation of a gallon of alcohol takes. Because this does not take place on the farm most of us do not consider it. It takes one BTU of energy to produce one BTU of gasoline. Because petroleum is so plentiful we do not concern ourselves with how much of it we burn to refine it to the form we want.

CHAPTER 3

GROWING ALCOHOL

DEFINITIONS

Air lock - A device through which gas can pass in only one direction.

Balling - The formation of large lumps by adding hot liquid to a starchy substance.

Enzymes - An organic compound that makes a chemical reaction take place more easily but does not react itself.

Fermentation - The breakdown of a complex organic molecule by a substance or organism.

Mash - The mixture of substrate and water that is fermented.

Liquefaction - Changing a solid to a liquid. In this process we liquify the carbohydrates in the grain.

PH - The degree of acidity or alkalinity of a solution.

Starch - A carbohydrate formed by the combination of several sugar molecules in a chain.

Substrate - The sugary or starchy product from which alcohol is made.

Sugar - Any of a number of basic carbohydrate molecules that have equal number of carbon and oxygen atoms and twice as many hydrogen atoms.

Yeast - A family of single celled fungi that ferments sugar and produces carbon dioxide and alcohol. (We us the genus *Saccromycetes.*)

     This chapter deals in detail with the growing of a crop of alcohol. Refer to the recipes and definitions to make the process more clear.

     In the following explanation we go through the production of a small batch using a grain substrate. The same basic procedure is used for other substrates, other than the preparation of the substrate, which will be covered later in the book.

Table III-1: Recipe For Grain

Tank Size	Grain	First Water	Taka-Therm	Second Water	Diazyme
50 gal	1.5 bu.	25 gal	1.5 oz.	12.5 gal	3 oz
100 gal	3.5 bu.	50 gal	3.5 oz.	25 gal.	7 oz.
200 gal	6.5 bu.	100 gal	6.5 oz.	50 gal	13 oz.
300 gal	10 bu	150 gal	10 oz.	75 gal	20 oz.
600 gal	20 bu.	300 gal	20 oz.	150 gal	40 oz.
1000 gal	40 bu.	500 gal.	40 oz.	250 gal.	80 oz.

GRINDING GRAIN

This can be done with any available equipment. You must, however, examine the grain to make sure all the kernels are broken. Any unbroken kernels will probably never be broken. If there is much dust or grain of flour consistency the grain should be introduced more slowly to the water. The dust will have more of a tendency to ball up and float on top even in cold water.

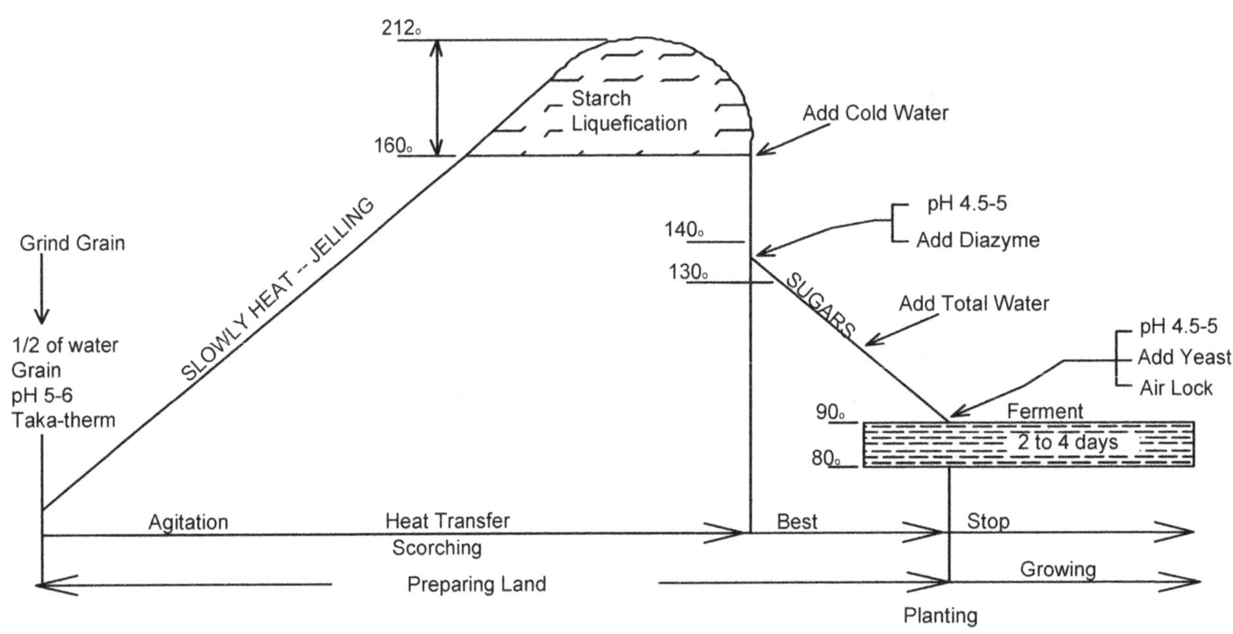

Figure III-1

MIX WATER WITH GRAIN

Heat 12 to 15 gallons of water per bushel to 190° F. Once you are in production you may use the coolant water as it comes out of the still for this purpose. It is already about 150 °F. so it will take less energy to cook the mash if you use this water. There is also hot water left in the boiler at the end of the distillation process but this water is quite acid and is best used for other things. Slowly add 1/4 to 1/3 of the grain, stirring as you do so. Add ½ ounce Taka-therm per bushel. After you have added the Taka-therm do not allow the mixture to cool below 180°F. until the starch is completely liquified or you will get starch retrogradation, which means the starch will go back together and no amount of boiling or enzymes will break it down. Slowly add the rest of the grain while heating and stirring the batch. Make sure the temperature does not fall below 180°. Heat to 190°F. and hold for 30 minutes. Always stir the mixture when you are heating it. Bring the grain mixture to a boil to make sure the grain is cooked. Allow to cool to 190°F. and add another ½ ounce of Taka-therm per bushel of grain. Hold at around 190° for at least another half hour. (Most batches will hold the heat well enough that these holding times are merely a matter of letting them sit.)

LIQUIFACTION

This term can be misleading if you expect a watery liquid. As the heat causes the starch to soften it will thicken the water until it is thick enough that the grain will be suspended in it. The Taka-therm will be turning the starch to sugar, which will make syrup, but it will still be thick syrup. If it is allowed to cool below 180° F. too soon, heated to boiling too quickly or the Taka-therm inactivated for any reason you will end up with a gel like cold oatmeal. When liquifaction is complete the mash will be somewhere between the consistency of heavy corn syrup and split pea soup.

COOL DOWN TO 140 TO 130 DEGREES

The next enzyme cannot be added until the temperature of the batch is below 140°. Add five to 15 gallons per bushel to cool the batch. The less water you add the more energy efficient the batch will be. You need to add a total of at least 17 gallons per bushel, however, or the yeast will start to die before it can turn all the sugar into alcohol.

ADJUST pH

Diazyme works best at a lower pH than Taka-therm, so you will need to test the pH. Pure water has a pH of 7.0. Grain is slightly acid and will lower the pH somewhat. The water you are using may not have a pH of exactly seven and the soil the grain was grown in makes some difference in the pH of the grain. Normally acid will need to be added at this time. Hydrochloric acid (also called muriatic acid) seems to work best, although any acid will work. Add the acid slowly and carefully until the pH of the batch is around 4.0.

ADD DIAZYME

Diazyme will convert the starches that the Taka-therm couldn't touch to sugar and further liquify the batch. Add two ounces of Diazyme per bushel of grain and let the batch sit for at least two hours. Unless the starches are completely converted alcohol production will be limited. We do have the hidden advantage that the yeast can produce its own enzyme but this slows up the operation and uses some of the energy that could have been converted to alcohol.

COOL DOWN TO 90 DEGREES

You can use cold outside air in the winter or cooling coils to cool the batch or just let it cool naturally on its own. Do not let it cool too long or the brew will have to be heated before adding the yeast so it will start working. In the winter get the temperature to just below 90°F. before adding the yeast so that cool air won't have a chance to cool the brew too much before the yeast has a chance to build up its own heat. In summer go down to 80° to keep the yeast from overheating the vat while it is working.

CHECK THE BATCH A LAST TIME

At this point everything is ready for the growth of the yeast. Yeast will grow best in an acid environment so check the pH one last time. It should still be around 4.0. You may also want to check for starch by adding a drop of iodine. Iodine turns blue when mixed with starch. Take a small sample, mix a drop of iodine with it and check the color of the drop. Some starch may still be present, but if your iodine darkens immediately, without any mixing, the conversion is not complete. After a few tests you will be able to determine by the consistency if you have good conversion.

ADDING THE YEAST

It is best to make a slurry of the yeast to make sure it is active and ready to go. To do this, place your yeast in warm water. NOT HOT just warm to the touch. You may put a little sugar in the slurry although we haven't really found this to be necessary. As soon as the yeast is actively foaming on top of the water stir it into the brew. Add around two ounces dry brewer's yeast per bushel of grain.

AIR LOCK

We now need to cut off the oxygen supply so that the fermentation process can take place. We don't want to grow yeast, we want to produce all the alcohol we can get. We do, however, have to make sure that any pressure built up by the carbon dioxide can get out or we can have problems. If there is a lid on your fermentation tank just have a sealable surface under the lid with a little weight on top. We have used a clear plastic cover held down by a big rubber band made from an inner tube. Sealed in this cover is a plastic hose through which the carbon dioxide can escape. The end of this hose is place in a bucket of water to keep oxygen from entering. This allows the metal lid to be lifted to check on the brew without letting oxygen in. The plastic expands as carbon dioxide builds up.

FERMENTATION

Figure III-2 shows what the yeast will tend to do to the temperature of the brew. It also shows the effect of different temperatures on the yeast. Above $100°$ the yeast begins to die. Below $80°$ it becomes less active. As the temperature rises the yeast becomes more active, producing more heat, until it gets to about $100°$. Then the yeast begins to die and the brew starts to cool off.

When the yeast begins to act the grain will form a thick layer over the top of the tank. This layer will fall when the action of the yeast slows down. There will not be any grain on top of the mash if the yeast has stopped for any reason. Killing the yeast will stop fermentation. Cooling the vat too much will stop fermentation. If fermentation stops because the batch got too cool it can be started up again by heating the batch. The biggest problem is that below $80°$ you will probably begin to produce vinegar. Vinegar producing organisms need some oxygen to grow. If the yeast had started to work, then vinegar producing organisms got started there will be deep bubbling with swirls coming to the top. If the brew has worked for at least a day and vinegar starts forming there is enough alcohol there to distill. There has to be alcohol there for the vinegar to get started.

The time it takes to complete fermentation will depend on several things.

    1. The amount of grain, or food supply, for the yeast.

    2. The alcohol content of the brew.

    3. The temperature that was maintained.

If you didn't prepare the brew well enough the yeast may run out of usable nutrients and will have to prepare them before digesting them. If you didn't put enough water in the alcohol content will get above 12% and the yeast will die. If the temperature is maintained near the $90°$ mark the action will be faster than if it is nearer the $80°$ mark.

One way to tell if you still have live yeast in the brew is to take some of the brew and mix it with warm sugar water. If there is foaming action the yeast is still alive and the brew can be reheated to ferment more if the grain is not all fermented yet.

CONCLUSIONS

This may seem like a complex process but it really takes more description than work. Most

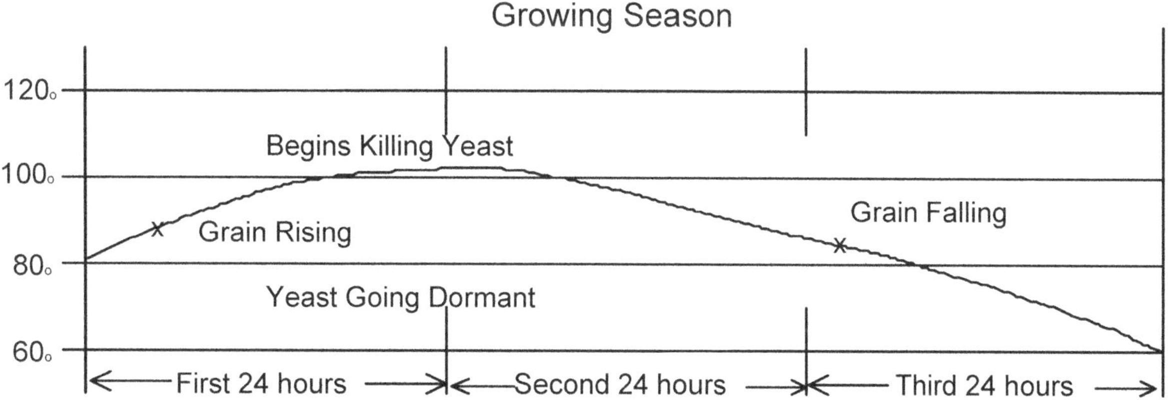

Figure III-2: The growing season for the alcohol crop. The line under the curve shows the temperature at which the yeast begins going dormant. The Curve shows the effect of the activity of the yeast on the temperature of the mash over time.

parts of the process are not too critical and leave room for modification. There is also a lot of waiting around and you can generally work the process into your other chores.

When grinding the grain, don't just grind the amount you need for that day, grind enough for a week or two at a time. Don't get too far ahead or there will be deterioration and loss, just as there would be with feed.

Heating and adding the Taka-therm can be very slow without any harm. Even the top temperature is flexible. As long as the brew is converted to a thick, creamy liquid you have done what is necessary. Cool down is the part of the process where most of the sugars are really made. The longer it takes the more complete the conversion. If it sits overnight and cools naturally, so what? Just remember that we want to end up with all the water and enzymes in a brew that has been liquified and is at a temperature of 80° to 90°. If it falls too low it will have to be warmed slightly.

Don't be afraid of the process or worry all through the preparation that something is going to go wrong. As long as there has been liquifaction and all the ingredients have been added you are in business. Just don't reheat the brew over 140° after you have added the Diazyme or above 100° while the yeast is working.

Large batches take longer to cool. You may wish to allow more time at the higher temperatures, in which case you could use less enzyme. The following table shows amounts of Taka-therm to add if the temperature is above 180° for varying lengths of time. These figures are calculated from assays of enzyme activity and do not take into account any loss of activity due to storage or for any variation in the cooking process. If you change the recipe be sure to check your mash and make sure it is smooth and creamy before continuing the process.

TABLE III-2

Time above 180°	Weight to Add per Bushel	Quantity to Add per Bushel
less than 1 hour	1.00 oz.	2 T.
2 hrs.	.50 oz	1 T
3 hrs.	.32 oz	2 tsp
4 hrs.	.25 oz	1 3/4 tsp
5 hrs.	.22 oz	1 ½ tsp
6 hrs	.10 oz.	1 tsp.

Figure III-3

Because the Diazyme continues to be slightly active all during the fermentation process you can

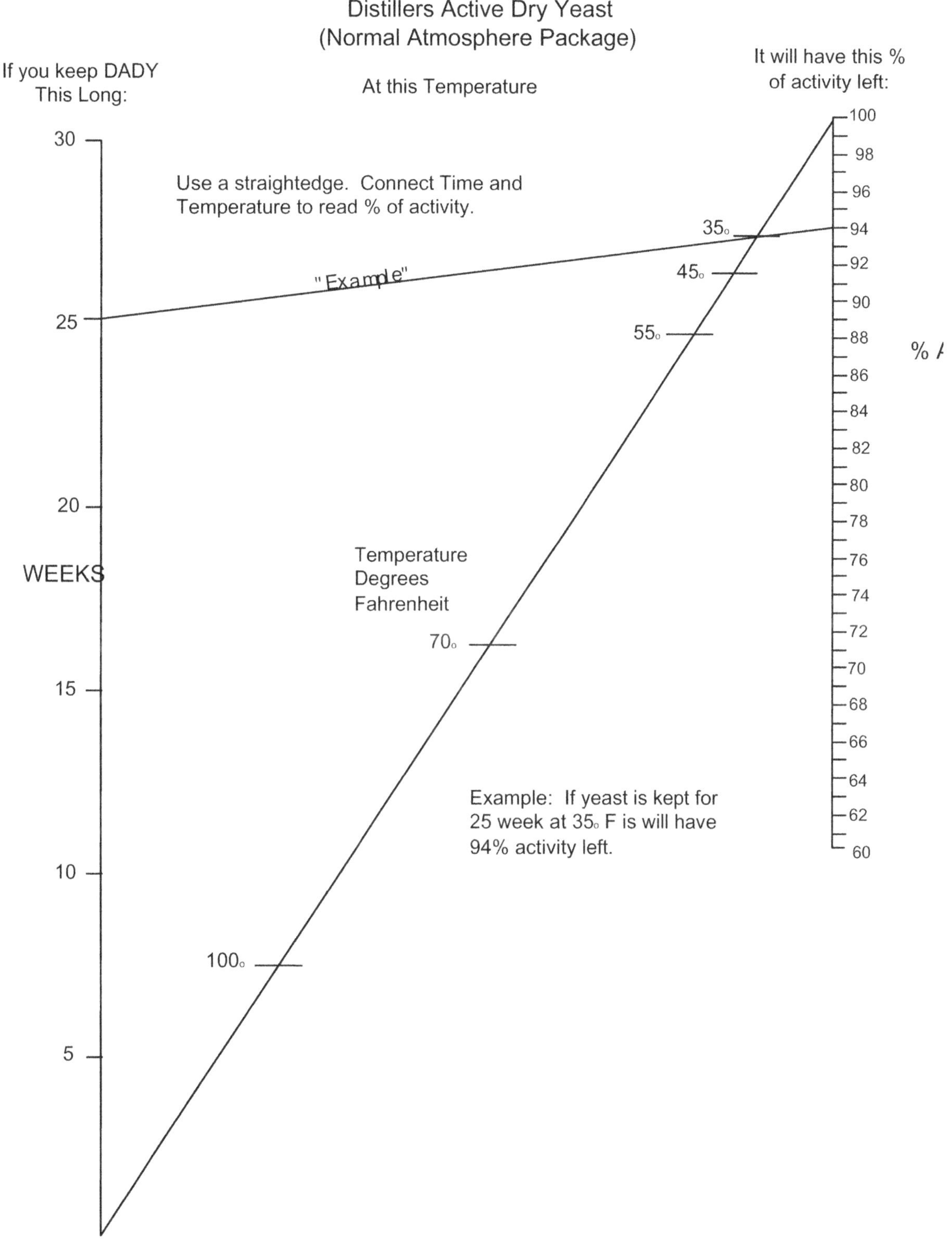

reduce the addition of Diazyme only if it takes a matter of days for your brew to cool down from 140° to 90°. A table of these changes is listed below.

TABLE III-3

Time above 90°	Weight to Add per Bushel	Quantity to Add per Bushel
to 2 hrs.	2.00 oz.	4 T.
2-24 hrs.	1.00 oz	2 T.
24-72 hrs.	.80 oz	1 2/3 T
72-130 hrs.	.60 oz.	1 1/3 T.

Another source of enzymes is Enzyme Development Corporation, 2 Penn Plaza, Suite 2439, New York, N.Y. They have enzymes in powdered form that are less likely to become denatured in storage than liquid enzymes.

OTHER RECIPES

Cane Sorghum Procedure

With cane products, including cane sorghum, sugar cane and several large stemmed grasses, there is much more fiber than sugar. Because the sugar is water soluble it will be carried with the juice if this juice is pressed out.

In the process of preparing such juice for molasses the juice is generally skimmed to remove fibrous particles floating on top. We tested these skimmings and found them to contain almost twice as much sugar as the juice that had been skimmed. Therefore we do not recommend skimming the juice.

Our research indicated how important it is to harvest cane sorghum at the proper time to get maximum yield from the crop. We have tested a number of varieties grown under different conditions and in different stages of development grown in both Kansas and California. The sugar content of the cane tested varied from 1% to almost 20%. In tests we performed in California, Sudan grass was discovered to have a higher sugar content than the cane sorghum, at least at the stage we tested.

We fermented some of the juice with higher sugar content. From 20 gallons of this juice we got 1.5 gallons of 170 proof alcohol. These fields yielded approximately 10 tons per acre from which 500 gallons of juice could be made. This would yield 32 gallons (pure alcohol) per acre. The procedure we used was to squeeze the juice from the cane using a press, bring it to a boil to sterilize it, cool to 90°F., add one ounce yeast per 40 gallons juice. Allow it to ferment until fermentation is complete - a shorter time than with grain in our tests - then distill.

No enzymes are necessary because the carbohydrates are already in the form of sugars and are thus available to the yeast.

Recipe for Potatoes

Shred potatoes, making sure all the juice gets into the cooking vat. Place the shredded potatoes in the vat with just enough water to keep the potatoes from scorching. (About 5 gallons per 100 pounds.) Add ½ ounce Taka-therm per 200 pounds potatoes. Heat to boiling while stirring. Stop heating and allow to cool to 190°. Add another ½ ounce of Taka-therm. Let sit for two hours. Add another 5 gallons of cool water per 100 pounds of potatoes. Allow to cool to 140° while stirring. Adjust the pH to 4.0. Add 2 ounces of Diazyme per 100 pounds potatoes. Cool to 90° F and add 1 ounce of yeast per 200 pounds of potatoes. Stir it in, stop stirring, air lock and ferment for two and one half days.

Another procedure for potatoes is to shred the potatoes, cover with boiling water and soak for two hours. Drain the water and repeat the procedure. Soak for a third time and press the water out of the potatoes. Collect all of the soaking water, add Taka-therm and Diazyme as above then add yeast and ferment. This procedure produces slightly less alcohol and because it uses more water is somewhat less energy efficient but you do not have such a sticky mess left in the fermentation vat as you do when you leave the insoluble solids in the vat during fermentation. It is a much neater process and the material remaining in the vat can be disposed of immediately after soaking.

To concentrate the starch a little you can use multiple vats. Pour the boiling water over potatoes that have already been soaked twice. After two hours drain that water into the vat that contains potatoes that have been soaked only once. Press all the water out, clean the first vat out and add fresh potatoes. After another two hours drain the water into the vat with fresh potatoes. After soaking two hours put the water into the fermentation tank.

Processing Sugar Beets

The carbohydrates in beets are in the form of sugar so no enzymatic process needs to take place. It is not as simple to remove the sugar from beets as it is from cane or fruits. Below are three methods used to remove the sugars.

Freezing

By freezing the beets until they are solid to the core the cell walls can be broken down and the sugar released. For alcohol production in the north, where thaws are rare, this method might prove practical for small scale production. The beets should be placed in batch size piles, preferably in an open, shaded area such as the north side of a building, and allowed to freeze solid. The frozen piles should then be covered with straw, leaves or some other insulating material to keep them frozen.

To use these frozen piles take them inside and place them in a large trough or tank to catch any liquid that drips while the beets are thawing. When the batch is completely thawed it should be smashed to squeeze out all the juice. The sugar will be in this juice, which should be boiled to sterilize it. Finally cool to 90° add 1 ounce of yeast for every 30 gallons of juice.

Of course this system only works in winter and in areas where it can be counted on that the piles will not thaw. If the piles do thaw before they can be pressed and fermented it will soon create a smelly, unpleasant mess. In other areas and times of year the soaking process can work. It is similar to the process for potatoes. Shred the beets into ribbons. They can be about a centimeter wide but should only be about a millimeter thick. Catch any juice that drips. Cover with boiling water and add sulfuric acid to adjust the pH to 4.0. Repeat the soaking three times to try to remove all the sugar. Press all the water out of the pulp after the third soaking. All the liquid should go into the fermentation tank, and the process proceeds as above with boiling and adding yeast. The multiple vat method described for potatoes can also be used here.

Cattail Recipe

The lower stem, crown and rhizome are the parts of the cattail where the starch is stored. If you cut these parts open you see a solid white core. This is the starchy portion. Chop or grind these parts very fine. Be sure to save all the juice with the ground portions.
Add enough water to be able to stir the mixture while you heat it. (About 10 gallons per 100 pounds.) Heat to 190°. Add ½ ounce Taka-therm per 100 pounds cattails. Bring to a boil. Cool to 190° and add another ½ ounce Taka-therm. Let sit for two hours, still stirring. Cool to 140° while stirring. If at any time the mixture gets too thick to stir, add more water, being careful not to over cool it. Adjust pH to 3.5 to 4.5. Add 2 ounces Diazyme per 100 pounds. Let sit two hours. Cool to 90°. Add 1 ounce distillers yeast per 100 pounds cattails. Stir and air lock.

The water content of cattails can vary widely depending on season, handling, etc., so add the amount of water you need to make a thick but stirable mixture.

Although we did not attempt to harvest the cattail roots at the time when they contained the maximum amount of starch our research showed that it would take about 300 pounds of cattail roots to make a gallon of alcohol. Harvesting, grinding and cooking would require large scale equipment. We did not try the soaking method described above for potatoes and sugar beets but that might work more effectively. Because of the per acre yield of cattails and their presence as a nuisance weed in many locations the numbers indicate it would quite likely be worthwhile to produce alcohol from cattail roots. Various reports cite the per acre yield at 70 to 140 tons. This would produce four times the alcohol per acre as corn, and would produce it on land that is not suitable for corn production but that would preserve habitat for many game and endangered species.

HYDROMETERS

One useful piece of equipment to understand in the process of making alcohol is the hydrometer. A hydrometer measures the density of water. Sugar in water makes it more dense. Alcohol in water makes it less dense. Alcohol with only a little water in it is even less dense. You can invest in a variety of hydrometers, because one hydrometer will not tell you how much sugar there is in your pre-fermented brew, how much alcohol there is in you fermented brew and how much water there is in your distilled alcohol. The only one you really need is the alcohol hydrometer to tell you how much water there is in your distilled alcohol. Because the pre-fermented brew is not pure sugar and water the reading of a sugar hydrometer will not be accurate. There are quite likely other things contributing to the density of the mixture. The same may be true for the fermented brew, although a beer hydrometer would be more accurate than a sugar hydrometer.

Another way of measuring the percent alcohol in your brew is by the boiling point. Alcohol has a lower boiling point than water and a mixture of alcohol and water will have a lower boiling point than water. How much lower will be determined by the amount of alcohol. To determine the boiling point place a pan of the mixture on the stove and heat it. Check the liquid temperature when bubbles start rising in the liquid. (not the temperature of the bottom of the pan. Don't let the thermometer rest on the bottom as you heat the mixture.) This temperature will rise for a short time then stabilize. The point at which it stabilizes is the boiling point. The graph on page 45 shows the boiling points of different concentrations of alcohol.

Of course, elevation and inaccuracy of the thermometer can also affect the boiling point of the liquid. To determine the accuracy of the graph check the boiling point of plain water with the same thermometer to see if water boils at 212°F.

AMERICA, GO FORTH AND BREW

CHAPTER 4

HEAT

DEFINITIONS

Boiling Point - The temperature at which a substance will change from a liquid to a gas.

BTU - British Thermal Unit - The amount of heat needed to raise one pound of water one degree Fahrenheit. Also, the amount of heat lost when lowering the temperature of one pound of water one degree Fahrenheit.

Condensation - Change from a gaseous state to a liquid.

Latent Heat - The amount of heat needed to change the state of a substance without changing the temperature.

Saturated Vapor - A vapor containing as much of a substance as it can at that temperature and pressure.

Specific Heat - The number of BTUs needed to raise one pound of a substance one degree Fahrenheit.

Superheated - Vapor heated above the saturation point so that a drop in the temperature will not cause condensation.

FACT SHEET

Mixture of Liquids - Two liquids mixed together act like a new substance with a new boiling point and freezing point.

Mixture of vapors - Vapors mixed together continue to act separately and retain the properties of the unmixed vapor.

Heat vs Temperature - Heat is the kinetic energy contained in a substance. Temperature is the property of a substance that determines if heat can be transferred. Heat is required to change the state of a substance (solid to liquid or liquid to gas) and the heat used to do that will not raise the temperature of the substance.

By now you should be able to prepare and ferment the mash or brew. Producing alcohol uses heat. We have a wide choice of fuels to use in this process. The more we know about heat the easier this choice becomes.

HEAT SUPPLIES

From Table IV-1 you can see that the heat produced by burning alcohol is a little more than half the heat produced by burning petroleum. This causes some problems but has some advantages.

Alcohol burns completely when it is burned. Petroleum products do not. You can see this easily by burning some of each. Smoke and fumes will be produced when the petroleum is burned, while the alcohol will produce no smoke and normally no fumes. You can test this with a carbon monoxide alarm, available at many hardware stores. The products of complete combustion are carbon dioxide and water. Carbon monoxide is produced whenever incomplete combustion takes place.

Because the only products of burning alcohol are the things you breathe out, alcohol heating devices do not need to be vented outside. This could produce problems only in a tight building where carbon dioxide could build up until it eliminates the oxygen. At this point you would be in danger of suffocation. The fire would also be in danger of suffocation and would probably go out before you do. Plants in the building would use the carbon dioxide and give off oxygen, eliminating that problem. There are also manufactured devices that could eliminate carbon dioxide from the vapor.

At this point we are looking for heat supplies to run our operation. Later we will look at alcohol as fuel for internal combustion engines. Here we will discuss it only as a heat supply. These are two different things. The internal combustion engine is not a heat engine. The steam engine is a heat engine. Under compression fuels will do more than just make heat. When a cannon goes off we are not concerned about how much heat it produces, we are concerned about the force behind the cannon ball. Likewise we are less concerned about the heat in the internal combustion engine than about the power behind the piston.

What we need in producing alcohol is direct heat. For this reason we can consider anything that produces heat. We must consider the cost, BTU content, availability and how it can be used in our system. In making alcohol we have seen that we need to add heat to produce the mash, then we lose the heat, then the yeast produces some heat. When we distill we will also add heat and lose heat. If we can increase the efficiency of all those processes, or use the loss from one to provide the heat for another, at least partially, we will be better off. Table IV-1 gives you a chance to compare various qualities of fuels that may be available to you. You can see from the table that we must burn more alcohol fuel, even utilizing all of its heat, to get the same amount of heat as many other fuels. The cost will depend on your production costs. As you will see later, the value of alcohol is much higher as a motor fuel than as a heat source. Solar and wind energy can be converted to heat and may be the heat sources most commonly used in the future.

UTILIZING HEAT

Every BTU of heat that our heat source produces must go somewhere. When we are cooking mash we want all of the heat to go into the mash. Any heat not going into the mash is considered lost. We can avoid wasting our fuel source if we can use this heat loss in some other way.

Heat must also leave the mash for it to get down to the fermentation temperature. The temperature is lowered when cooler water is added because the heat in the mash passes into the water. Thus only a little of the heat is lost.

TABLE IV-1 PROPERTIES OF HEAT SOURCES

Fuel source	Fuel Content	% Burning Efficiency	Conversion Factor (Fuel Content x Efficiency)	Representative Local Cost	Cost /Million BTUs
1 gal #2 oil	140,000 BTU/gal	80[1]	112,000 BTU/gal	$1.24/gal	$11.25
1 gal #2 oil	140,000 BTU/gal	70[2]	98,000 BTU/gal	$1.24/gal	$12.87
1 gal. ethanol	75,670 BTU/gal	70[1]	52,969 BTU/gal	$ .60 to $1.85/gal	$11.33 to $34.93
1 gal ethanol	75,670 BTU/gal	100[6]	75,670 BTU/gal	$ .60 to $1.85/gal	$7.93 to $24.45
1 cu. ft. Natural gas	1,000 BTU/cu.ft.	80[1]	800 BTU/cu.ft.	$.071/cu.ft.	$8.44
1 cu. ft. Natural gas	1000 BTU/cu.ft.	70[2]	700 BTU/cu.ft.	$.071/cu. ft.	$10.11
1 therm LP gas	9,300 BTU/therm	80[1]	7,440 BTU/therm	$.17/therm	$22.88
1 therm LP gas	9,300 BTU/therm	70[2]	6,510 BTU/therm	$.17/therm	$26.11
1 kWh Electric heat	3,413 BTU/kWh	95[3]	3,242 BTU/kWh	$.09/kWh	$27.76
1 kWh Electric heat	3,413 BTU/kWh	80[2]	2,730 BTU/kWh	$.09/kWh	$32.97
1 cord Hickory	35,300,000 BTU/cord	70[4]	24,710,000 BTU/cord	$125/cord	$5.06

Cost per million BTUs can be recalculated using local fuel costs using the following formula (Local fuel cost x 1,000,000 BTUs)/conversion factor.

(1) high efficiency furnace   (2) standard furnace, maintained   (3) baseboard or fan units   (4) woodburning furnace, wood less than 12% moisture   (6) utilizing vent duct heat

We also found that when the yeast was working it produced heat. Though this is a low temperature heat it can be dissipated into the room to assist in maintaining a constant temperature in the environment for other fermenting vats that may not be producing as much heat.

Obviously all of the above heat loss can be used to heat the fermenting room. If this process takes place in a shop or other building it would make other winter chores more pleasant. It could even be in a basement or other part of the house that is not used for other purposes and you could use the excess heat to heat the house. Many animals will gain or produce more and new born babies survive better in a warmer environment, so the heat could be piped into a chicken house, farrowing house, lambing quarters or barn and again increase the profitability of the farm.

As we get into distillation you will see that high temperature heat losses are necessary for the effective operation of the still. REMEMBER, IF WE UTILIZE ALL HEAT LOSSES THE COST OF PRODUCTION IS GREATLY REDUCED. Just using the spent grain to feed cattle reduces or eliminates the cost of the grain for alcohol production. Using the heat, after the work of producing the alcohol is finished, in an area where it would be needed anyway, allows us to share the fuel costs with another operation. Any time we can use a byproduct of the production of alcohol for another purpose the cost of producing the alcohol gets closer to zero.

HEAT TRANSFER

Heat is transferred only from warmer areas to cooler areas. There are three forms of heat transfer; conduction, convection and radiation.

Conduction is the passing of heat from molecule to molecule. In conduction heat is transferred to the adjacent molecule containing the least amount of heat. In the cooking and fermentation process energy is conducted through the metal of the boiler, through the liquid next to the metal, through the metal to the top of the boiler and through the air outside the boiler. Heat does not travel very far through conduction before it is dissipated. Conducted heat travels at different rates through different substances.

The thicker the metal of the boiler the more heat is conducted through the metal to be radiated to the air rather than the brew. You want heat to be transferred to the brew so you should use the thinnest metal possible that will withstand the heat of the fire.

Convection is the transfer of heat by the movement of the fluid (air or water) containing it. When the heat has been conducted through the boiler and into the liquid inside the liquid becomes less dense and rises, carrying the heat with it. The air outside the boiler will also rise. The more movement of air and water the more movement of heat through convection. Convection is what cools the outside of the simple stripping column after heat is conducted through the metal and radiated to the air around it. Heating by convection is a slow process and does not produce very even heat. Heat must be radiated from the moving fluid to the thing it is heating. The more swiftly the fluid is moving the less chance there is for this to happen.

Radiation is the transfer of heat from one substance to another through waves. These waves of energy travel until they strike a molecule. This is the way heat is transferred from the fire to the metal, the metal to the water and from the metal to the air around the boiler and still. Radiant heat can travel great distances through space but in the dense atmosphere of the earth it does not travel very far and in the dense liquid of the cooker and boiler it travels even shorter distances. Without convection the air or water molecules immediately around the metal would soon reach the same temperature and heat transfer would stop. The more air movement around the metal the more cool molecules move into the area so that more heat can be transferred. That is one reason you should stir the mash while you are cooking and while boiling.

Heating a building with the hot water from the still is a relatively straightforward and simple application. But what if you have animals that need protein all year round and you do not have any use for the heat in the summer?

You could use the hot water for air conditioning, particularly if you are operating more than one column. Refrigeration units called absorption units are manufactured to air condition spaces. One of the smallest is made by Arkla Industries. It will operate with intake temperatures of 175° to 205°F. It's rated capacity is three tons, which it achieves at water temperatures of 195°. The water outlet temperature is 45°.

The unit consists of tubing containing lithium bromide solution with a high pressure side and a low pressure side. This tubing is in contact with the water at three points. The hot water heats the solution on the high pressure side. It is then released to the low pressure side, where it is vaporized and absorbs more heat, cooling 55° water to 45°. This chilled water can then be used to air condition a house, hog parlor, etc. The lithium bromide solution is then pumped to the high pressure side and gives up the heat it absorbed to the condenser water, which enters at 85° and exits at 95°. The rate of hot water flow must be four to 22 gallons per minute. The condenser water can be recycled by allowing it to air cool. The unit uses a maximum of 25 watts of electricity so operation for 24 hours would use a maximum of eight kilowatts of electricity. The unit itself is 68' tall, 29" wide and 28" deep. For more information write Arkla Industries, P.O. Box 534, Evansville, IN 47704 and ask about their model WF36 Absorption chiller.

Excess heat from the process can also be used to dry the distiller's grain for later use.

CHAPTER 5

DISTILLING ALCOHOL

You have already made all of the alcohol you are going to get out of this batch. The still does not make alcohol any more than a combine makes grain. The still separates the water and alcohol just as the combine separates the chaff and grain.

The liquid in your mash tank is now beer. If conditions have been just right it is very strong beer. You can test whether you have made a good batch, a poor batch or no alcohol at all by smelling the brew. If it smells like beer you have alcohol. If it has an acidy-sweet smell there was a problem somewhere in your process. (There are other smells that also indicate a problem and many of them are quite unpleasant.) A good batch will have more bite than a not so good batch. With only a little practice you will be able to distinguish a good batch.

If you would like to know more accurately whether you have a good batch or a bad batch you could get a beer hydrometer. Put it in your mash before you add the yeast and record where it floats. When you have finished fermenting check the brew again. If the beer hydrometer sinks farther into the fluid you have made some alcohol. The percentage reading may not accurately reflect the alcohol content of your batch because of other things present in the solution.

You may also wish to do a sugar test at this point to determine if fermentation is complete. A simple way to do this is to use a product called Tes-tape, a urine sugar test available at drug stores. Take one drop of your brew, add nine drops of water and compare the color with the chart. Take the percentage on the table and multiply by 10. You should not have any sugar left after fermentation.

If you have a good strong beer and have a low yield when you distill, then you can start looking for a problem in the still. Otherwise review the fermentation process.

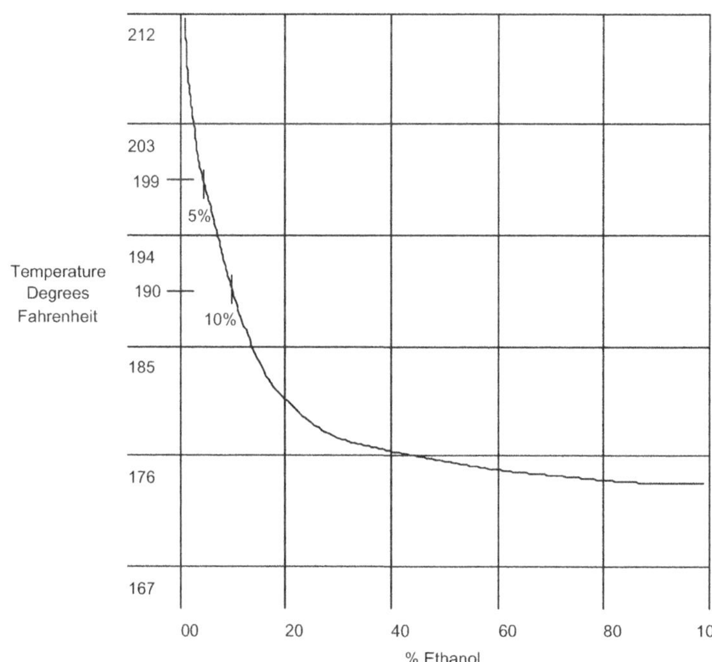

Figure V-1: Boiling point of liquid with various concentrations of alcohol and water.

STILLS AND COMBINES

If the alcohol is a crop, the still is the combine that harvests it. Combines work because chaff is lighter than grain. Stills work because water condenses at a higher temperature than alcohol. In order to understand what is taking place inside the still you must understand the behavior of vapors. Review Table V-1 and the definitions and facts accompanying it.

HOW A STILL WORKS

We can see an interesting thing when we boil water on the stove. No matter how much heat we apply the temperature never gets higher. The steam only comes off faster. The liquid is using all of the heat to change to a vapor rather than get hotter. The temperature of the vapor can rise if more heat is applied to it. This is true for any liquid. The amount of heat the liquid needs to become a vapor is the latent heat of vaporization and is different for different liquids. As you can see from Table V-1 alcohol takes less heat to vaporize than water. It also boils at a lower temperature. The same amount of heat is released when a vapor is condensed as is absorbed when it vaporizes and it condenses at the same temperature.

Table V - 1 PROPERTIES OF ALCOHOL AND WATER

Property	Ethanol	Water
Weight Lb./gal.	6.7	8.3
Specific Gravity	0.794	1.0
Density (lb./cu.ft.)	49.3	62.4
Boiling point (°F at 1 ATM)	173.0	212.0
Freezing point (°F)	-173.0	32.0
Specific Heat (BTU/lb./°F)	0.60	1.0
Latent Heat of Vaporization (BTU/lb.)	396.0	970.0

When two liquids are mixed together the boiling point of the mixture is a point between the boiling points of the two liquids. Just where this boiling point is depends on the concentration of the components of new liquid. Because we are starting with a brew containing less than 15% alcohol the boiling point will be closer to that of water than alcohol. The vapor, as it comes off the liquid, will contain the same percentage of alcohol as the liquid. What we want to do is concentrate the alcohol by removing the water.

If the temperature of a vapor is below its boiling point the saturation point is much lower than if it is at its boiling point. If the temperature of a vapor is above its boiling point it is superheated. Vapors mixed together act separately. Both vapors will be at the boiling point of the liquid, which means that the water vapor is below its boiling point and the alcohol vapor is superheated. The water vapor is carrying more heat than the alcohol because of its higher latent heat of vaporization. Some of the water vapor begins to condense immediately because the air above the boiling liquid is saturated.

What we want to do is cool the vapors slightly so that the water condenses and the alcohol does not. When we condense the water the heat it releases has to go somewhere. The larger the mass and the higher specific heat of a substance the more heat it can absorb. In order to absorb heat the substance has to be at a lower temperature than the condensing vapors. The more slowly we cool the vapors and the more carefully we control the temperature the more water we can remove before we condense the alcohol. Because some water will still be present in the vapor at 173° when the alcohol condenses and because 100% pure alcohol attracts some water we cannot produce pure alcohol. However, alcohol with some water in it burns fine and is even cooler and safer, so the water is not a problem.

WHAT DOES THIS MEAN

In the still we first boil the brew. We then have saturated water vapor, superheated alcohol vapor and the mass of the still. At first even slight cooling condenses a great deal of water.

As the water condenses the alcohol vapors become more concentrated, cooler and closer to their saturation point. When this happens some of the alcohol will begin to condense with the water vapor. Thus little drops of an alcohol and water mixture will be falling through the vapor that is rising through the still. The alcohol is more concentrated in this liquid than it was in the brew so the boiling point is lower. When the droplets reach the point where the temperature is equal to their boiling point they evaporate again and the vapor travels upward. The heat of vaporization comes from some of the water vapor, which then condenses and falls through the still. As distillation proceeds the alcohol vapor goes higher and higher in the still and gets more and more concentrated. The idea behind any still design is to get the alcohol to the desired concentration before the temperature reaches the condensation point of the alcohol.

It is very easy to make a still. All you need to do is boil the brew and cool the vapor slowly. Finding the right balance of heat addition, heat loss and mass to be able to control the proof is a little more difficult.

One word of warning here. It is very easy, if you do not follow certain safety precautions, to make a still that will blow up in your face. When you add heat to a boiler make sure you have a way for the vapor to escape if the normal vapor outlet gets clogged. Adding heat will build up pressure. If there is no release for this pressure it will continue to build up until it is released. If you keep adding heat the pressure will release itself by bursting your boiler. Most boilers are strong enough to hold a great deal of pressure. That pressure will be released with a great deal of force. Also, alcohol vapor is explosive. It would be useless as a fuel if this were not so. It is quite safe to handle in the liquid state but if vapors from the still come in contact with a spark they will explode. Alcohol does not explode like gasoline. It burns more slowly and goes off with a whoosh rather than a bang, and it is more likely to ignite something else. To avoid this make sure the still is well made and take care of it while it is in operation. The boiler should have a pressure relief valve that will release at relatively low pressure.

The boiling and condensing process takes place in different ways in different stills. In the next chapter we describe what happens inside various common stills. These descriptions should make the process more clear.

CHAPTER 6

STILLS AND COLUMNS

SIMPLE MOONSHINE STILL

We are all familiar with this type of still, having seen it in pictures or jokes. Few know what goes on inside it to make it work. Even the old moonshiner probably didn't know what was happening. He just knew that if he put his brew in the boiler (which was usually just a clean old barrel) the vapors produced would condense in a coil of copper tubing that passed through another barrel filled with water. Sometimes this barrel, or condenser, was just a creek flowing by the site. It was discovered that putting a coil of tubing over the top of the boiler would increase the proof of the alcohol greatly. All kinds of things were put in the coil to get the proof up high enough that good

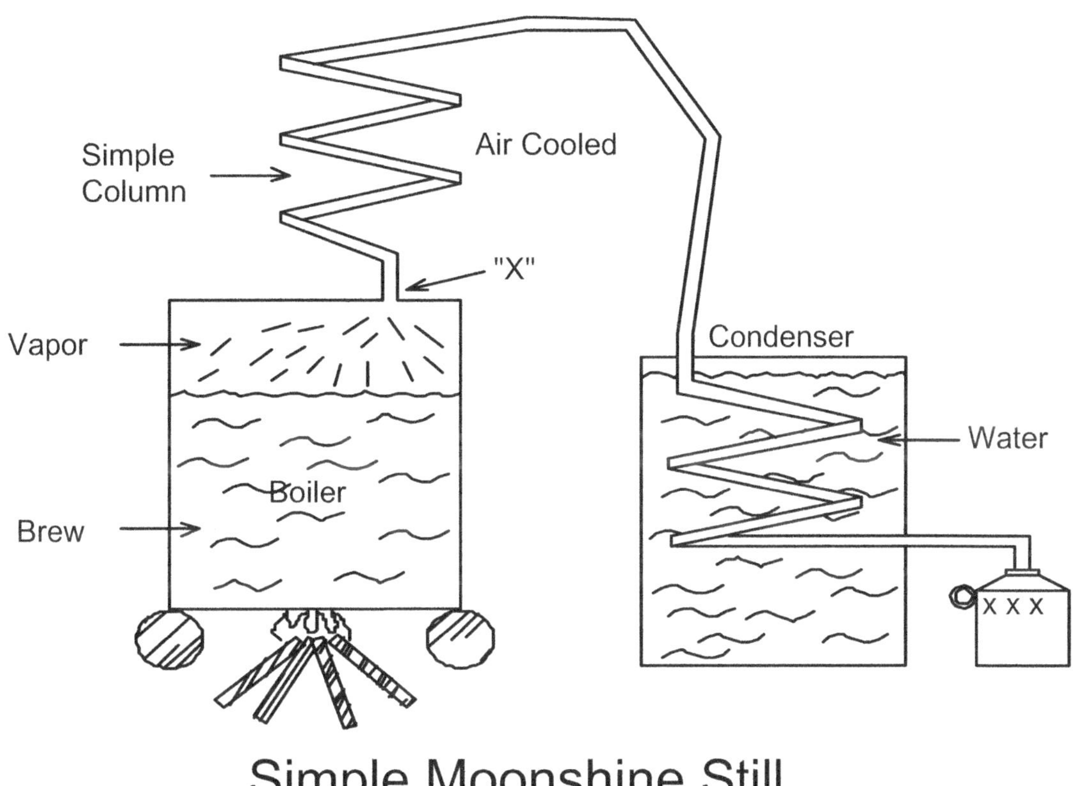

Simple Moonshine Still

moonshine could be produced in only one pass through the still. It was not the copper tubing in the condenser that produced the good alcohol. Rather it was the tubing or whatever was between the boiler and condenser. Lets follow what happens inside that still.

The brew in the boiler boils off in a vapor of mainly alcohol and water. This passes through the tubing on top of the boiler. As the heat held in the vapor passes through the copper and into the air around the tubing the water and some of the alcohol are condensed and run back into the boiler. The vapors that did not condense pass on and are completely condensed by losing enough heat to the water in the barrel or condenser. The alcohol simply drips out of the end of the tubing into a jug. The end of this tubing had a valve or restriction in it in some stills to keep the vapor in the tubing saturated so that the water would condense. Copper is the most effective material to use because it allows heat to pass through faster than most other materials. It also does not leave traces of poisonous heavy metals in the moonshine. Most of the old boys who owned the stills did not know what was happening inside the stills, they just knew that one setup made better booze than others. The coil on top of the barrel is really a stripping column. When we discuss other types of columns you will see more clearly how it works.

The more fire under the boiler the faster the vapors will come off. If the vapors are coming too fast to be cooled by the air passing around the stripper coil the proof will go down. Also, there might be so many vapors that the condenser could not keep up. With a simple still the stripping action is at the mercy of the temperature and wind speed of the air passing around it. If we want a constant proof for our fuel we have to do something about this. The old still has everything needed to make drinking alcohol but the proof is not high enough to use as fuel after the brew has passed through. Therefore, we have to beef the old feller up.

DOUBLER COLUMN

The doubler column was probably discovered by some old moonshiner messing around with his still. He found that if the vapor was passed through a pool of liquid about half the water stayed behind in the pool. The vapor line from the boiler is hooked to the doubler at X and the vapor passes into the bottom pool. The pool absorbs heat from the vapors so that the temperature of the pool rises above the boiling point of alcohol. At that point the alcohol vapor passes through the pool along with about half the water vapor. The remaining water vapor stays behind in the pool. The vapor rises to the next pool in the column and the same thing happens, all the way to the top of the column. This is an infinitive process so that eventually removing half the water does not make all that much difference. At this point the vapors are allowed to pass into the condenser.

The water that is left passes out through traps that control the pool depth. There can very well be some alcohol left in the runoff. One of the disadvantages of this column is that the percentage of alcohol in the runoff can be quite high. This runoff can be rerun through the same column or run through another column to retrieve the remaining alcohol.

This type of column is used a great deal in commercial distilleries and is quite big. The diameter of the vapor pipes and pools determines the hourly capacity of the column. Enough heat must be applied to the boiler to drive the vapor all the way through the still. Too small a heat supply would not be able to get the vapor to the top of the column before the temperature of the vapor dropped so low that all the alcohol would condense into one of the upper pools. The diameter of the column must match the capacity of the heat supply

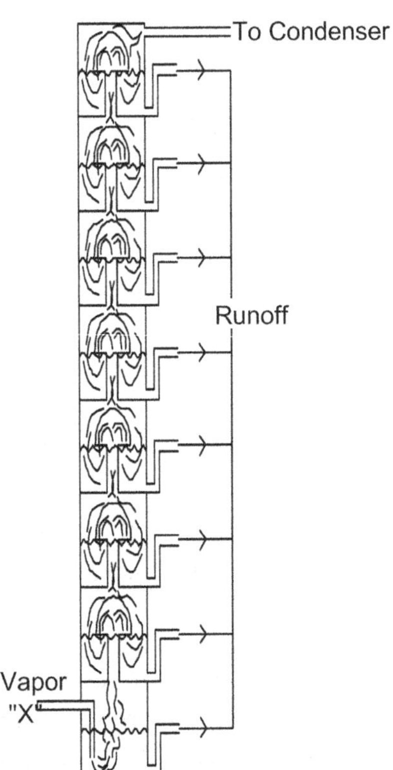

Doubler Column

of the boiler. The height or number of pools that the vapor passes through will determine the proof of the alcohol that passes out through the condenser. This is the key to any distilling column.

In any still the gallons per hour are determined by the heat supplied to the boiler. The proof is determined by how effectively the water is stripped out of the vapors and how effectively other contaminants are removed by in the column.

MARBLE COLUMN

The marble column is a complex column that has been used for many years. It is not normally used commercially but operates on the same principle as many commercial types.

This column attaches to the boiler in the same way the Doubler did only the vapors pass through a pile of marbles and heat them up. After the marbles they encounter a coil with cooling water and condensation takes place. The liquid drips off and begins to fall back through the marbles. This cools the marbles from the top down. With the vapors heating the marbles and the condensate cooling them down all the marbles are above the boiling temperature of the alcohol but are gradually cooler higher in the still. When the condensed liquid reaches the place where the marbles are the temperature of its boiling point it will evaporate again and travel higher in the still. This action very effectively takes the water out of the vapor and washes out any other unwanted compounds.

Once the vapors get past the stripping coil they are free to pass out of the still to the condenser. We can measure the temperature of the vapor as it passes into the condenser. By setting the flow rate of the water in the stripper coil so that it absorbs just enough heat to cool the vapors to a predetermined temperature we can adjust or maintain the percentage of water that is stripped out.

This is probably the simplest type of column to control. The old simple moonshine still had no control at all. If this column were hooked to the boiler in place of the coil of tubing the proof of the drinking alcohol could be accurately set. It would make really hot white lightening. This was and is where this simple, easily controlled column is used.

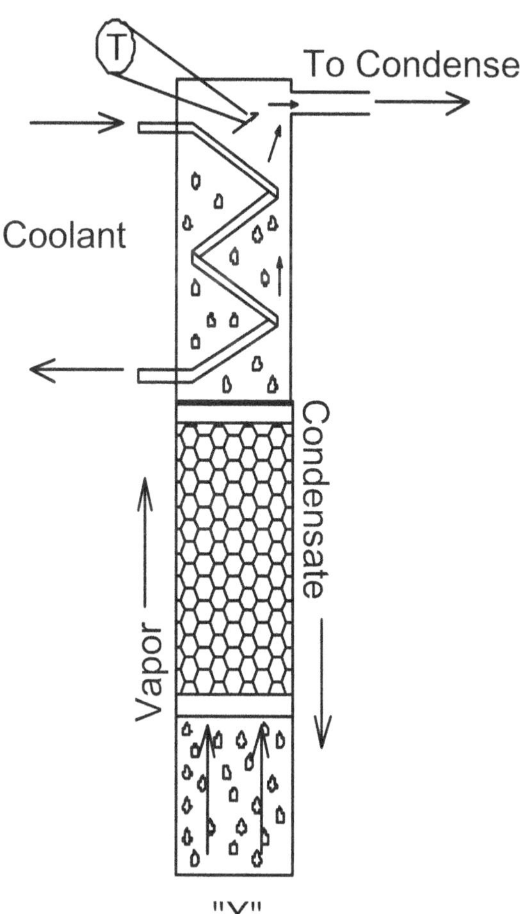

STEAM COLUMN

This is definitely a commercial column. It has many plates inside in place of the marbles, although the action is the same. However, instead of the boiler being under the column and containing the brew it contains only water, or there is a separate steam generator that supplies heat to the plates in the column (dotted lines). Either way the steam heats the plates. The brew is pumped into the column near the top, flows over the plates and falls through the holes. Each plate becomes a small boiler working on a thin film of brew. Vapor quickly comes off, just as in any boiler. This vapor and the steam coming up and the brew and condensate coming down produce the same vaporization and condensation process as described before.

This type of column very effectively separates the water and alcohol, although it is much harder to control. Also, this is the first column that requires a pump to get the brew into it. These columns are much higher than any we have discussed before, especially the marble column. This is because the brew enters the column cold and must be heated by the steam as it falls through the plates.

REMOTE POST COLUMN

This column was designed with the first section of a Doubler column with a marble column sitting on top of it. It uses the action of each type of column where it is most effective, along with a section of simple stripper column. The vapor enters from the boiler at X, although it can be set farther away from the boiler as runoff will be caught in a separate container rather than run back into the boiler. An automatic modulating valve can be placed in the probe position to control the flow of

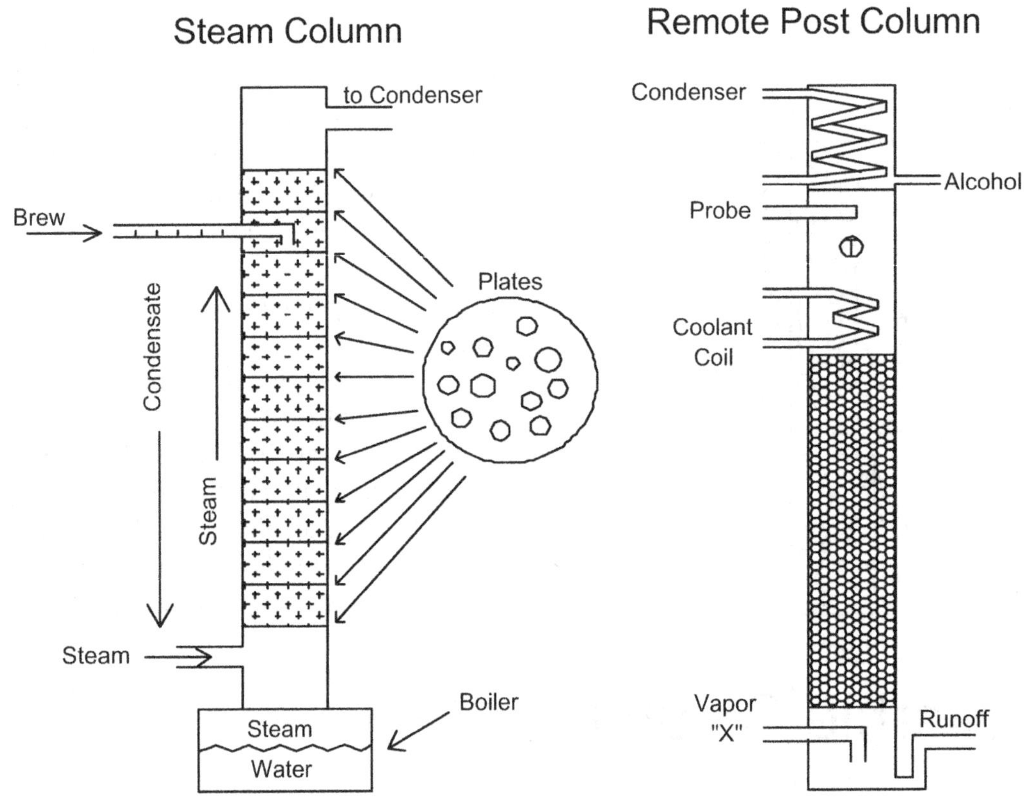

water through the stripper coil. This will accurately control the proof of the alcohol without constant monitoring. The condenser is simply set on top of the column as most of the heat will be taken from the vapors through the striper coil, leaving very little heat for the condenser to absorb.

The name comes from the facts that the column can be set up at some distance from the boiler and it looks like a fencepost. The height of the column is greatly reduced because the Doubler takes out half the water right at the start. In the Marble column all the water vapor had to pass through the still. A small volume of water will absorb a great deal of heat (because of its high specific heat) and that is precisely what we want to happen when the vapor is at its hottest and contains the most water. After that the Doubler becomes less effective so we go to the marble column. We can hook the column up so that simple stripping similar to what took place in the moonshine still takes place before the vapor even reaches the Doubler.

EXAMPLE: Lets say that we use the old moonshine still with enough plumbing between the boiler and the column that the proof as it enters the column is 100. That is 50% water vapor. At the Doubler half of this water vapor would be taken out, leaving 25% water. If we did nothing else to the vapor we would have 75% alcohol vapor or 150 proof. There are a lot of small stills that will not get the proof that high, even running it through a couple of times. By combining still types we have increased the capacity many times over what a still of either type of similar size could produce.

As the condensed liquid runs out the bottom of the column we can catch it and decide if any further processing would be worthwhile. You may add it to your next batch or save it until you accumulate enough to make a batch. It is a relatively clean liquid that is very acidic. Saving it and running a batch of runoff may not give you much alcohol but is very effective at cleaning the boiler.

OVERVIEW AND CONCLUSIONS

Figure VI-1 shows the changes in temperature of the vapor as it comes to a boil then as it goes through the still. A strong batch of brew will start boiling at a temperature of around $190°$. As it boils some of the water will condense and drop back into the mixture in the boiler immediately. Because of this the boiler temperature will gradually increase and the relative amount of alcohol in the vapor will decrease. When the boiler temperature reaches $212°$, or whatever the boiling point is at your elevation, turn off the boiler because all the alcohol has been removed.

The vapor that reaches the still will gradually decrease in temperature until the alcohol condenses. The boiling point of pure alcohol is $173°$ so you should have the valve set on the stripping coil at a slightly higher temperature so you do not condense all the alcohol and send it back down through the column. At the beginning of the run alcohol will come out of the Remote Post Column, or any batch still, faster than at the end. In some stills the volume will remain the same but the proof will drop. Because the proof is held constant in the Remote Post Column the volume of liquid will be reduced. Production will be relatively constant until the boiling temperature reaches about $207°$ then will drop until almost no alcohol is produced at $211°$.

Now you know how to produce alcohol. The fermentation process produces the alcohol. The distillation process separates the alcohol from the water. The next step is to try it on your own. As you experiment with it you will develop the best procedure for your particular situation.

CAUTION - Again, remember that there is the possibility of explosion and fire. A roaring fire under a homemade boiler can cause pressure to build very rapidly. Be sure to have some sort of pressure relief that will blow off if your vapor line gets clogged. Alcohol vapors will also blow up if ignited. Be sure the vapors from the still can be exhausted away from the boiler fire.

Figure VI-6
PRESSURE IN THE STILL

To test the pressure in the still you can use the carbon dioxide hose. Leave the hose open so that vapor can escape through it during operation of the still. The vapor will follow the path of least

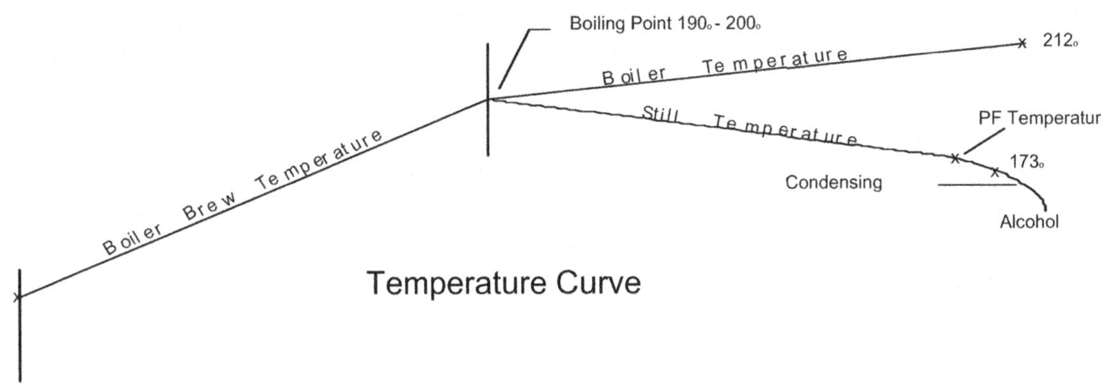

resistance. By putting the carbon dioxide hose under water until the vapor passes through the still then measuring the depth of the hose under the water you can determine the pressure in the still. Twenty-seven inches of water equals one pound per square inch pressure. We were surprised to learn that we operated our still with the hose only 20 inches under water or at .8psi.

Once we tested the pressure in this manner we decided to use the carbon dioxide hose as our pressure relief valve. We have no shutoff on the hose and keep it in 21 inches of water. The carbon dioxide bubbles out freely when we close off the column. When we open the valve to the column and turn on the fire the vapor will pass through to the column unless it is somehow blocked and pressure begins to build. The vapor will then be released through the carbon dioxide hose before enough pressure can be built up to cause any damage. The vapor will also be released into the water where most if it will condense, rather than into the air where the alcohol vapor could be somewhat irritating to the lungs.

CHAPTER 7

EQUIPMENT

Up to now we have discussed the process of making alcohol without describing the equipment needed in detail. You know that you have to cook the mash, ferment it then boil the alcohol out into a still. We know how the still has to work to get the job done. Now we will discuss the materials you will need to get started.

TANKS

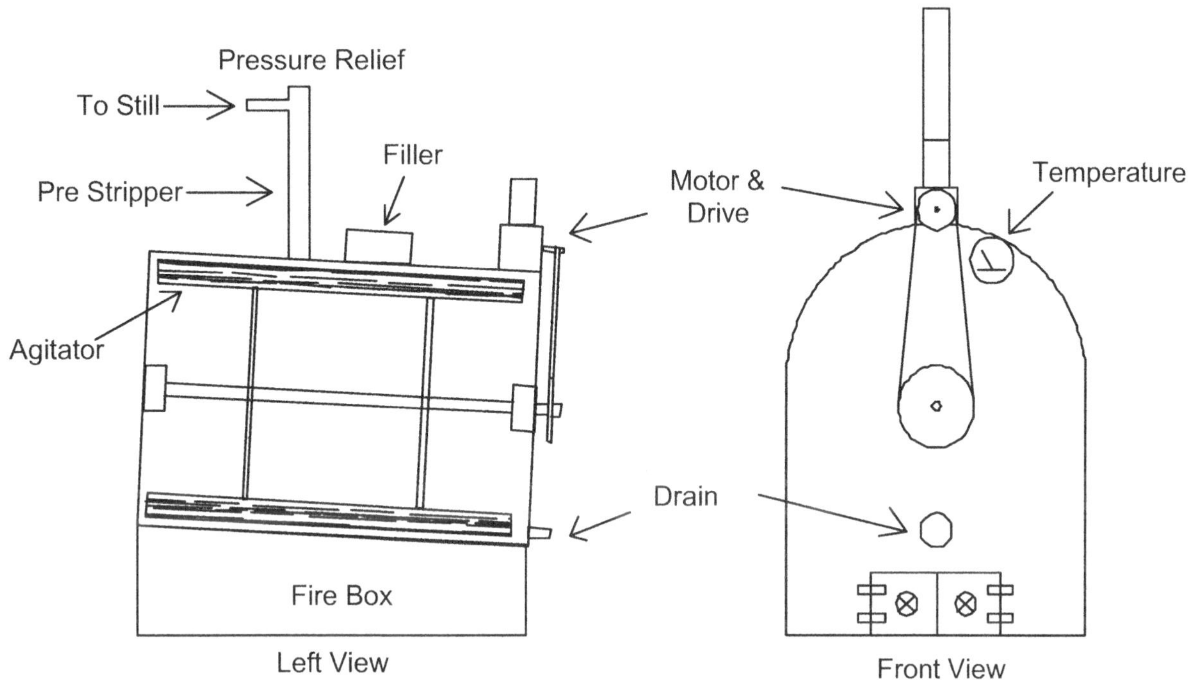

Cooker-Fermenter-Boiler

Cooker - This can be anything from a bucket to a barrel to a large vat. It is recommended that you start small in a bucket or 55 gallon barrel before going on to full scale production, just to get the feel of how to produce alcohol. Choose your fuel supply, set the barrel on top and get started. You can stir by hand for a while. Learn what is happening by watching it closely. If you are using a sugar-based substrate like cane or sugar beets you do not need a cooker.

Fermenter - Again, this can be anything from a bucket to a barrel to a large vat. It can be the same container as the cooker or it can be a different container.

Boiler - The easiest boiler to produce is an old oil barrel. It can be sealed up tight and can be heated either on the end of on the side. With a single burner heating it on the end will give more even heat. With two or more burners or with a burner like that from a gas oven heating on the side is better. This gives more surface area to heat as well as more evaporative area in the tank.

Most oil barrels have a 2" bung and a 3/4" bung. The 2" has normal pipe threads but has no tapper to tighten up on. You will have to provide a collar to tighten down on to make a seal This is easily done with a close nipple or a conduit nut screwed into the bung. This will be your fill hole and vapor outlet. Put a 3/4" valve into the small bung to drain out the brew when you have finished distilling. Bore a hole in the top of the boiler and insert a rubber plug. Into this plug insert a probe type thermometer. This will be where you check the temperature and will also be your pressure relief.

With this type of simple boiler you will need to strain the distiller's grain out of the brew before putting it into the boiler. A big old funnel, milk strainer or something similar lined with window screen can serve this purpose. The grain will still be quite wet after straining. It can be dried somewhat by bouncing the grain on the screen or by squeezing it, either by hand or in a press. The grain will be quite spongy if it has been well cooked and fermented.

When you think you are ready you can produce a combination vat. If you compare the temperature curves of cooking and distillation you will see that the same temperatures are required. Both require stirring if the grain is in the tank. Heat transfer will be more rapid in both, even if grain is not present, if they are stirred. You can see that the same tank could be used for both processes.

For fermentation you want an air tight tank with no stirring. You also need a pressure relief to let the carbon dioxide out. The boiler needs to be air tight with a pressure relief, too. So why not build a tank that can contain the whole process. Below we outline two ways to approach this.

Direct heat tank - The drawing on the previous page shows a round tank lying on its side with a firebox under it. The fire box is enclosed to reduce heat loss. The whole tank is pitched forward just slightly to allow for complete drainage. The drain is large to allow the distillers grain to come out after the process is complete. The firebox is vented through a chimney in back and has dampers in the fuel doors to control the burn. A very important part of this system is an agitator placed inside that will slowly stir the mash. The best speed is 11 rpms or slower. The exact best speed would depend on the substrate and the number of paddles on the agitator. The agitator should be built

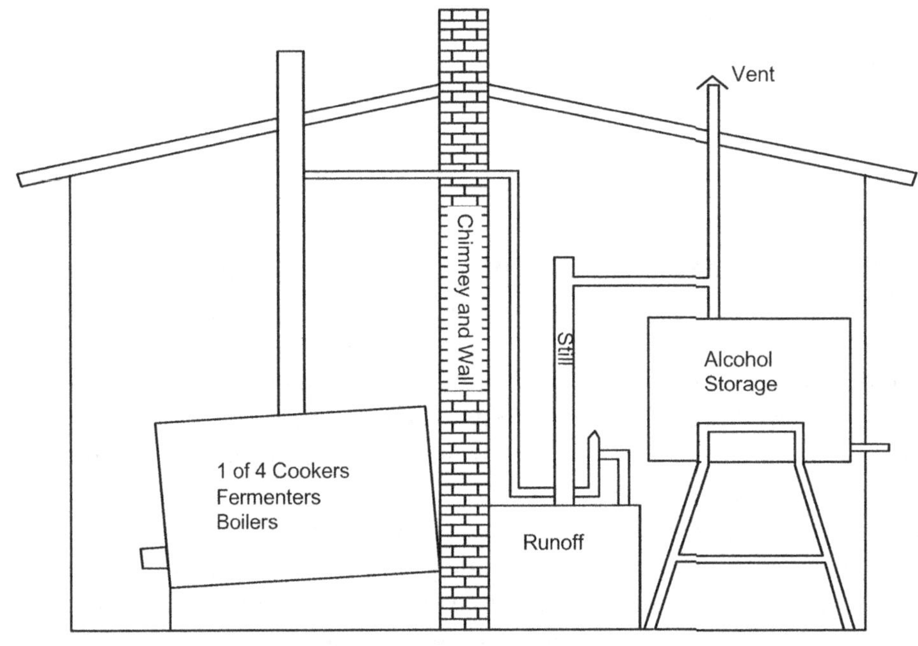

Typical Farm Fuel Plant

something like a combine reel with angle iron at the outer edge with very little clearance between the angle iron and the edge of the tank. A packing or carbon-faced seal like those used in pumps should be used to seal the point where the shaft goes through the tank. A speed reduction box is mounted on top with a drive motor. The agitator is chain driven. Belts would get hot and deteriorate too quickly. Provide some sort of slip release so that if the agitator gets stuck for some reason the agitator will not break or the motor burn up.

The vapor pipe is best placed near the center of the top to let the vapors out easily. A hatch, which can be sealed down, should be near the front so you can fill the tank. Water can be pumped directly into the tank. The temperature can be read by putting a regular boiler thermometer near the top of the front of the tank. If you like you may add other conveniences like a liquid level gauge or a small tap valve somewhere near the middle to draw off samples for pH tests and other purposes.

Indirect heat tank - This tank is like the direct heat tank except that it has a water jacket under the tank. The firebox can still be there to heat the water. That makes the tank like a double boiler that you would use to heat milk or make frosting. There is no way scorching can occur with the water jacket. Agitation would be needed only to mix ingredients and speed heat transfer but the speed and side clearance of the agitator would not be so critical.

This type of tank would cost considerably more but has many advantages. It could utilize solar and wind power to preheat the water jacket. The water can be heated either under the tank or separately. Thus you can use any heat source and transport the heat to the tank The tank can also be cooled by passing cool water through the water jacket. This might be necessary in the summer during fermentation in many locations. You could take the heat from the water jacket and utilize it in other locations when the process is complete or take the distilled mash to another tank to heat the water jacket of that tank, then use that heat to cook the next batch.

With three vats of either sort a near continuous operation can be set up. The coolant water from the still, which is around $200°$, is almost hot enough to cook the grain. This water can be added slowly to the tank with the first 1/4 of the grain and the first addition of Taka-therm. The agitator in the tank should already be working to assure even cooking. When all of the grain and a sufficient amount of hot water are added the water can be diverted elsewhere (home heat, domestic hot water, etc.)and the gelatinization and liquifaction allowed to take place. After this has taken place cold water is added and the mash stirred until it is cool enough to add the Diazyme. Then it is stirred and cooled until it is cool enough for the yeast. With large tanks this may be a long time. During fermentation the tanks should be stirred briefly to cool it slightly and make sure fermentation is taking place evenly in the tank. After fermentation is complete the agitator and fire are again started up and the brew distilled. The tank is then emptied, the mash strained and the liquid put into a storage tank to cool. The runoff from the still can also be put n this storage tank. This is the water that will be used to cool the tank before the Diazyme is added because it is already quite acidic. If the mash is strained immediately after the fermentation process the distiller grain will dry quickly because it is quite hot.

REMOTE POST SUPER COLUMN

During the last several years of his life Robert Brautigam worked to perfect the Remote Post Column. The first Remote Post Column was a piece of the downspout of his house with marbles inside. With the passing of time it became more sophisticated until it was one of the most effective, efficient small stills available. They are no longer being manufactured but they are not difficult to make. Directions can be found in Appendix A.

SAFETY

For safety reasons it would be wise to equip your system with an automatic shut off valve that will stop the flow of fuel to the system if the temperature of the boiler exceeds 212°. This would be effective on systems using liquid or gaseous fuel. Solid fuel fires can be controlled by adjusting the flow of air. This can be done thermostatically, although wood, coal or biomass fires generally need to be watched more closely than gas or liquid fires and the air could be restricted manually. In addition to the thermostat there should be a five-pound pop-off valve, another precaution against explosions

STRAINERS AND DRIERS

When you get into larger production you will want to reduce the amount of handling required for the distillers grain. The tanks should be high enough to allow for gravity flow into the screening area. Otherwise you will have to pump the mash or put a tank underground to catch the liquid.

Industrial strainers can be purchased or an old combine sieve can be made to work. An old fanning mill, or something similar would work. To squeeze out the last or the water something like a bailer or an auger with a restricted end could be used. The goal is to get the meal dry enough to feed.

You might want to dry it so as to mix it more easily with your feed ration, store it or even sell it. This can be done in several ways. You will need to squeeze most of the water out, apply some heat and provide some air flow. You could use the hot water from the distillation process or build a solar drier. Once the distiller's grain has been squeezed out fairly well it can also be held for several weeks in air tight storage.

WATER STORAGE

When you drain the mash you will have a great deal of water left over. This water can be stored, cooled and used to cool the mash before adding the Diazyme. The runoff from the doubler portion of the still can also be added to the water storage. Both are quite acidic and will lower the pH so the Diazyme will be more active.

Even after straining, the liquid will have small particles of very nutritious food in it. These will settle out in the storage tank. The bottom of the storage tank can be drained and fed to chickens, hogs or put into a fish pond.

TEMPERATURE

60°	65	70°	75°	80°	85°	90°	95°	100°
80%	79.0%	78.5%	77.5%	76.5%	75.5%	75.0%	74.0%	73.0%
81	80	79.5	78.5	77.5	77	76	75	74
82	81	80.5	79.5	78.5	78	77	76	75
83	82	81.5	80.5	79.5	79	78	77	76
84	83	82.5	81.5	80.5	80	79	78	77.5
85	84	83.5	82.5	82	81	80	79	78.5
86	85	84.5	83.5	83	82	81	80.5	79.5
87	86	85.5	84.5	84	83	82	81.5	80.5
88	87	86.5	85.5	85	84	83.5	82.5	81.5
89	88	87.5	86.5	86	85	84.5	83.5	82.5
90	89.5	88.5	88	87	86	85.5	84.5	84
91	90.5	89.5	89	88	87.5	86.5	86	85
92	91.5	90.5	90	89	88.5	87.5	87	86
93	92.5	91.5	91	90	89.5	89	88	87
94	93.5	92.5	92	91.5	90.5	90	89	88.5
95	94.5	93.5	93	92.5	91.5	91	90.5	89.5
96	95.5	95	94	93.5	93	92	91.5	90.5
97	96.5	96	95	94.5	94	93.5	92.5	92
98	97.5	97	96.5	95.5	95	94.5	94	93
99	98.5	98	97.5	97	96.5	95.5	95	94.5
100	99.5	99	98.5	98	97	97	96.5	96
160	158	157	155	153	151	150	148	146
162	160	159	157	155	154	152	150	148
164	162	161	159	157	156	154	152	150
166	164	163	161	159	158	156	154	152
168	166	165	163	161	160	158	156	155
170	168	167	165	164	162	160	158	157
172	170	169	167	166	164	162	160	159
174	172	171	169	168	166	164	163	160
176	174	173	171	170	168	167	165	163
178	176	175	173	172	170	169	167	165
180	179	177	176	174	172	171	169	168
182	181	179	178	176	175	173	172	170
184	183	181	180	178	177	175	174	172
186	185	183	182	180	179	178	176	174
188	187	185	184	183	181	180	178	177
190	189	187	186	185	183	182	181	179
192	191	190	188	187	186	184	183	181
194	193	192	190	189	188	187	185	184
196	195	194	193	191	190	189	188	186
198	197	196	195	194	193	191	190	189
200	199	198	197	196	195	194	193	192

Table VII-1: Hydrometer Corrections

HYDROMETERS

An alcohol hydrometer can be used to measure the proof of the alcohol. Only an alcohol hydrometer will work. Battery, antifreeze or beer hydrometers will sink to the bottom. The proof and percentage scales on your hydrometer are calibrated to be accurate at 60°F. Above 60° the glass the hydrometer is made from will expand, changing the scale, and the liquid will be less dense, making the hydrometer float at a different level.

The table on the previous page gives the correction for both proof and percentage at different temperatures. To read the table check the temperature of your alcohol. Find that temperature at the top of the table and the reading on the hydrometer in the left hand column. Follow the line of your reading until you are under the temperature of your alcohol. That is the corrected reading. For example, if your alcohol is 95° and the proof reads 188 the true proof is 178.

THERMOMETERS

The best type is a dial thermometer. They read like a clock and have a nail-like probe. These can be found in hardware and grocery stores and are used for roasting meat in microwave ovens. You will need at least two, one for the boiler and one for the still.

With a little ingenuity you can come up with all sorts of handy-dandy little work savers as you learn to make alcohol.

OTHER MEASURING EQUIPMENT

You don't need to set up a laboratory, but you might want to get a better idea of what you are doing. You will find that with a little experience you will be able to look at and smell the mash and tell if it is acidic or if the liquifaction is good. You will also be able to tell how quickly fermentation is taking place. If you want to make more accurate tests there is a variety of equipment on the market for starch tests, pH tests, sugar content tests and others. Any high school chemistry teacher would be able to describe these tests to you and any scientific instrument sales person would be happy to sell the equipment to you.

If two people measure the same thing with two different pieces of equipment they can come up with widely different readings. This is true because of two factors, instrument accuracy and instrument tolerance.

Instrument tolerance is how far off the true reading an instrument can be. The thermometers we generally use can be 2% off. That means that when you read 212° it can be between 208° and 216°. This is expressed as 212°± 4°. Hydrometers, at 60° allow for a tolerance of 2%. That means 160 proof alcohol is 160± 3. The narrower the tolerance, as a general rule, the more expensive the piece of equipment.

The accuracy is how closely the instrument can be read. Generally accuracy is one half the markings on the piece of equipment. For instance, if there is a mark for every two degrees on the thermometer and the needle points between two marks the reading is the odd number between the number of the two marks. For instance, halfway between 100° and 102° is 101°. You cannot say that the thermometer reads 101.2° because you cannot read tenths of a degree on a thermometer with markings every two degrees. If there was a mark for every degree you could tell if the temperature was 101.5° but could not differentiate between 101.4° and 101.6°.

How does this affect your proof readings? Lets say you have alcohol ready for testing. The thermometer seems to say exactly 96°F. and the proof reads exactly 190. With a possible error of 2% on each instrument the temperature reading is 96°±2°, or between 94° and 98°. The proof is 190±4 or between 186 and 194. When we check the table we find that 186 proof at 94° is 176 proof; 186 proof at 98° is 174; 194 proof at 94° is 185 and 194 proof at 98° is 184. So our proof could be anywhere between 174 and 185 and we cannot get a more accurate reading with these instruments.

More accurate instruments are available with tolerance of as little as .2%. With these instruments you would be able to check proof within one proof point in the above test and, if tested at 60° the proof would be 190.0±.4. However you would need one hydrometer to test proofs from 145 to 165 , one for proofs from 165 to 185 and one for proofs above 185. Since anything above 140 proof can be used for fuel it is not that important to have a more accurate reading.

HEAT SUPPLIES

We have discussed heat and hopefully you have a better understanding of the subject now. Here we will discuss a variety of fuels that you could use to heat your system. For production of the necessary amount of alcohol you will be looking for a great deal of heat.

Field Residues - This fuel source has some interesting possibilities and also some drawbacks. A ton of straw, corn or milo stalks has a very large volume. If you choose to burn them as your heat source you will be handling a great amount of mass in the system. Trying to put a ton of straw under the boiler could be a real adventure. This problem could be handled easily with an indirect heat boiler. A pit could be made with water pipes passing over the top in such a way that almost all the heat produced could be picked up. This heat could then be carried to the boiler in the water.

Another thing that must be considered is what will happen to the fields and air if this material is stripped from the ground and burned. It would be foolish to reduce the fertility of the soil and limit the yield of crops just to get fuel for the alcohol production process. It is up to each individual to decide on their farming practices and fuel source but this one doesn't pencil out on a farm that is working for self-sufficiency.

Utilities Fuel - This cost is probably easiest to figure out. There is a definite price per kilowatt hour, cubic foot, therm, etc. You can sit down with a representative of these utilities and quickly make a rough estimate of the fuel cost for a gallon of alcohol. First figure the amount of fuel it would take to cook the mash (specific heat x number of pounds x degrees temperature increase) then distill the mash (latent heat x number of pounds) as though it were water. This will give you a higher figure than what it will actually take but if each fuel supply is figured the same way it will be a good basis for comparison.

Alcohol fuel - This could be your fuel. Use your own alcohol to fire the system if your cost to produce the alcohol is close to half the cost of other fuels. Why must the cost of the alcohol be half that of other fuels if alcohol is such a good fuel? We will discuss that in detail in the next chapter but in short, the alcohol has fewer BTUs per pound than petroleum fuels. You will see that to heat our homes we will gain a great deal by not having to vent the furnaces. In an internal combustion engine alcohol has an advantage over other fuels. But for direct heat where we cannot utilize the heat of the combustion gasses the alcohol has no BTU advantages.

Methane gas - For some farmers this would be the perfect fuel. Normally you need to have something to feed the distillers grain to. If these animals are in confinement the fuel is being made for you. The composition of methane is very close to that of alcohol except that it is always a gas at normal temperatures. It is difficult to store in large amounts and should be used as fast as it is produced. Thus it suits this operation very nicely. You might say you are getting two fuels in one process. It doesn't work quite that way but nearly so. With this system the animals are being raised for profit (hopefully) and in the process we get fuel for the farm for nothing.

The drawback to this fuel source is that the number of animals it will take to produce all of the fuel needs will be greater than the number of animals that can get all of their protein needs from the distillers grain. It will take three times as many animals to produce the methane than are needed to consume the distiller's grain. Making methane is as complex a field as making alcohol and there is not room to go into all the considerations here.

Solar Heat - This is another great possibility. The only problem is that the sun doesn't shine all the time and we must always have a backup. A combination of solar and methane might be ideal. There are many possibilities for the use of solar energy in the process, although it is more efficient at lower temperatures than it is at high temperatures. We have cooked the mash with a solar collector and have operated the still on it but the collector was not designed for the high temperatures required for the distillation process.

Wind Heat - Again, the wind doesn't blow all the time so the heat source requires a backup. Perhaps a combination of wind and solar generated electricity would be a good option. Small scale electric generators are becoming more cost effective all the time, both because of the development of new technology and the rising cost of power.

Using wind directly as a fuel supply is quite simple. A DC generator is turned by the wind. A bank of heaters equal to the capacity of the generator is attached to the generator. There are no storage batteries. When the wind is blowing the electricity is converted directly to heat.

CHAPTER 8

ALCOHOL FUEL USES

FIRE

First, lets consider what is necessary to create fire. There must by oxygen and a fuel to be oxidized. Oxidation of iron, commonly called rust, is a slow form of fire. Oxygen is uniting with iron and very slowly burning it up without any visible flame. At the other extreme there is gunpowder, which can unite with oxygen so fast it explodes.

We all know that if gasoline is ignited it burns very rapidly but it must have oxygen to unite with. By controlling the oxygen, or air, we can control the burn. With alcohol the same thing happens but it burns slightly more slowly. When you burn gasoline you will notice that there is smoke. This means that not all the fuel is completely burned. Alcohol fires produce no smoke. It is one of the few fuels which, once ignited, will burn completely with no smoke and leaving no residue behind. Smoke contains carbon monoxide because the burning of the carbon source is incomplete. Alcohol produces no poisonous fumes when burned, just carbon dioxide and water vapor.

HEATING

Alcohol can be used to great advantage for comfortable home heating. Due to the fact that it burns clean it doesn't require venting. Thus with alcohol fuel we can benefit from all the heat it produces. This is not so with any other fuels we use. Once a chimney is placed inside a home to vent unburned fumes there is a hole in the house. The normal flow of hot air is up. Therefore we are not only sending unburned fuel up this hole, we are losing a lot of the air inside the home that we have heated.

Using alcohol as a heating fuel the heater can be vented directly into the house. Not only can all the heat be utilized, the air does not dry out as it does with regular furnaces. An alcohol heater will hold winter humidity at a very comfortable level. The only problem would be the carbon dioxide. There is a possibility that the carbon dioxide would build up in the house, replacing the oxygen we need to breathe. This would happen only in a very tight house and there are two ways to eliminate the problem. The simplest is to have plants in the house. Plants use some of the carbon dioxide and give off oxygen just as we use oxygen and exhale carbon dioxide.

If you want to use alcohol to heat something else, such as a tank of water, or wish to cook with it we cannot use it any more efficiently than any other fuel, so we have to figure our costs the same way we would with these other fuels. This is why we said earlier that using alcohol to heat the mash would not be economical unless you can produce alcohol for half the cost of the other fuels.

To use alcohol for home heat you can burn it in any oil burning device. The weight of alcohol is near that of kerosene or diesel fuel. The jets or floats in these devices will work nearly as well on alcohol as the fuel they were designed for. There may be a problem with some of the pumps in forced oil systems. Some use plastic parts that could be dissolved by the alcohol. Some may need a little oil mixed with the fuel for lubrication. Use 5% to 10% vegetable oil mixed with the alcohol. Remember, we still have to supply that alcohol with air so there has to be air passage to and from the fire. We cannot seal up the vent where it comes from the furnace or the gasses released by burning will build up and create pressure, stopping the oxygen from getting to the fire.

Fueling the internal combustion engine is the reason most people want to make alcohol. It is one renewable fuel that can easily replace gasoline or diesel fuel.

We said earlier that internal combustion engines are not heat engines. Some people would have you believe that they are. That way they can use the lower BTU content of alcohol to discourage its use. Just as there is a difference between heating a tank and heating the air there is a big difference when alcohol is put into an engine. REMEMBER, WE ARE USING ENGINES DESIGNED FOR GASOLINE OR DIESEL FUEL. The ideal engine for alcohol would be a cross between a diesel engine and a gasoline engine. But we can, for now, get by with some minor modifications on existing engines.

We said that when we are heating air has to be supplied to the fuel to keep it burning and it has to be vented to make room for more air. An entirely different thing happens in the internal combustion engine. The air is mixed with the fuel on entering a sealed chamber, the mixture is compressed and then ignited. The greater the compression of the fuel the more power there is from the resulting explosion. Gasoline can only be compressed only so much before it ignites from the heat of compression. Alcohol ignites at a much higher temperature and can be compressed much more than gasoline. Also, the speed with which gasoline burns makes it difficult to burn all the fuel inside the cylinder. The burn is over almost as soon as it starts. The alcohol really doesn't explode and will burn all the fuel even if it has to keep burning in the exhaust pipe. The alcohol also contains some water, which expands a great deal when it is heated and helps the engine stay cooler, resulting in longer engine life. These are the main factors that have to be taken into account when using alcohol fuel. However, until an alcohol engine is on the market in this country we cannot take full advantage of these factors and will have to make do with modifying gasoline and diesel engines.

First, lets get GASOHOL out of our minds. A proof of more than 196 is required when alcohol is mixed with gasoline or the two fuels will separate. The theoretical maximum of any still is 197 proof. The Remote Post Column is very good but to make alcohol of that high proof consistently would be out of the question. It would take another step and another run to get pure alcohol. Besides, gasohol is mostly gasoline anyway and we want to get away from a reliance on gasoline.

With gasohol out of the way lets see what it would take to run our engines on straight booze. The first thing that needs to be done is correct the air mixture. Either we change the air flow or the fuel flow. When we look at a carburetor we see that the speed of an engine is controlled by the air flow so we had better leave that alone. So the fuel flow must be increased to be more in balance with the air flow. This means boring out the fuel jets. This kills two birds with one stone. It allows the thicker alcohol fuel to get through the small hole and it gets the air to fuel ratio closer to the optimum.

Before tearing up the carburetor make sure the engine is running well on gasoline. If not, get it fixed first, then after you modify the carburetor you can try to get the gasoline engine to run nearly as well on alcohol. There will be some differences in the way it runs.

If you are not a precision mechanic, take the carburetor to a mechanic who is. These parts are delicate and this is not a job for a hammer and pliers mechanic. Taking it to a professional with the following instructions could save you a lot of money and cursing.

The first thing you need to do is drill the carburetor jets out about 40%. Check tables to find the size of your jet and multiply that by 1.40. Use the drill bit closest to that size to drill out the jets.

While you are there, take the float and place it in some gasoline. Mark how deep it sits, then put it in some alcohol (the proof you are going to run in the engine.) Add weight to the float to make it sink to the same depth. Remember - alcohol is heavier than gasoline and the more water in the alcohol the heavier it is.

Now, with the jet bored out and the float adjusted, wash the whole thing in some alcohol. Gasoline is a cleaner but alcohol will clean things out that gasoline won't touch and will even clean out deposits left by gasoline. Also, drain the gas tank and wash it and the lines with alcohol. Get several fuel filters and ask your mechanic to show you how to change them if you do not already know how. You will need to replace them as things begin to get really cleaned up.

Put the carburetor back on just the way it was and see if the old gal will start up. If it is cold it probably won't. In this case shoot some starter fluid into the air intake and try again. Once the engine warms up things will begin to settle down. Remember that a true alcohol engine would have a high compression like a diesel engine but you don't have that with this old gas buggy.

You will probably find that the engine won't idle smoothly. Try to adjust the idle jet. If you can't get the engine to idle smoothly you may have to bore it out slightly, too. Even then it may not idle smoothly. Don't get discouraged, we aren't finished yet. Let the engine warm up and run it a while to remove carbon deposits left by unburned gasoline in the cylinders. It should begin to settle down and start to run pretty good.

We said that alcohol burns slower so loosen up the distributor hold down and move it back and forth to get a better timer setting. It should be set somewhat ahead of that for gasoline. You can tell if it is set right by listening for knocking. A stethoscope will help you eliminate knocking completely. Do not make this adjustment until the engine has run long enough to clean out the carbon deposits. (This will be as soon as the engine warms up thoroughly.) These adjustments will work wonders on some engines but on others they won't do much of anything.

Next look at the intake manifold. It is probably sweating like mad if there is not a strip of tin around it and the exhaust manifold. If there isn't one it will help a great deal to make one. Simply wrap some tin around the outside of both the exhaust manifold and intake manifold to allow the exhaust to heat the fuel slightly as it enters the carburetor. It may also help to put a metal hose on the air intake of the air cleaner and pick up air from around the exhaust manifold, where it is hotter. You can also heat the fuel line with some of the coolant water or move it closer to the engine. With gasoline there is a danger of vapor lock if you do this but alcohol vaporizes at much higher temperatures.

If your first try using the new fuel is an old tractor you may not have to bore the jet out. Most tractors have adjustable jets and it may open up enough to work. If it doesn't, bore out only 20% and try not adjusting the float. If you can get by with this the engine can be made to run on either gasoline or alcohol. The timing can be adjusted and marked for each fuel. Wrapping the exhaust and intake manifold will not generally do any harm with gasoline. In fact, it works even better.

Now that you have done all of that, what problems will you have using alcohol in an engine built for gasoline? If your jets are too small it could burn your valves. This is not as much of a problem in cars that are designed to use unleaded gasoline because they have hardened valve seats. You could install stainless steel racing valves to handle this possibility. If the jets are too large you will be burning too much fuel. Alcohol will not flood an engine like gasoline will. It will keep right on burning, sending flames out of the tail pipe if there is enough of it. That is not a safe driving practice. You may have problems starting in cold weather. This can be taken care of with starter fluid.

These are the main things that can be done without tearing down the engine and rebuilding it to increase the compression.

DIESEL ENGINES

It is probably best to just leave these alone. There are conversion kits available that were developed as long as 20 years ago, but when the price of fuel went down relative to other prices in the 1980's most people stopped working on the development of alternatives and little has been done in the past several years.

The M&W gear Company in Gibson City, IL developed a conversion kit in the early 1980s for turbocharged tractors. The system consists of a tank containing 100 proof alcohol, a pressure line and a feed line. When the engine is idling no alcohol enters the engine. As the turbocharger boost pressure increases it forces air through the pressure line, pressurizing the tank and forcing alcohol out. The higher the boost pressure the more alcohol is injected into the engine. This injection is controlled by the size of the orifice on the feed line. A different size orifice is used with each different engine.

The advantages of using the injection system are that the airstream is cooled, providing a denser charge of air for the condenser, the burn is cooler and more powerful, the exhaust is cooler and the cylinder sleeves, pistons and valves stay cleaner and last longer. The lower combustion temperatures mean less heat is transferred to the cooling water and oil. There is more alcohol in the system the higher the load, resulting in the cleanest burns when the blowby is greatest, meaning the oil stays cleaner. The greatest advantage is that there is more power with less fuel.

In one test a tractor produced 125 horsepower at rated speed on diesel fuel alone while consuming eight and one half gallons fuel per hour. When the alcohol was injected the engine produced more horsepower so was throttled back to 125 horsepower for comparison. The engine consumed six gallons of diesel fuel, one gallon of alcohol and one gallon of water in an hour. The total fuel was reduced by ½ gallon, one gallon of that was water, which is considerably cheaper than diesel fuel, one gallon was alcohol, which is easily and inexpensively produced on the farm. The amount of diesel fuel per hour was reduced by 2 ½ gallons.

Even when you have installed a system like this the tractor is still dependent on some diesel fuel so this is not the ultimate answer.

The biggest word of caution is that the compression of a diesel engine is at the point where alcohol has the greatest power. Because of its high octane rating alcohol is capable of producing more power than gasoline and much more than diesel fuel when utilized properly. The horsepower of the engine increases dramatically and most diesel engines are not equipped to take it.

Even pure alcohol will not mix and stay mixed with diesel fuel. Castor oil can be mixed with alcohol but that is too expensive to use as fuel. Sunflower oil will mix but has a tendency to separate over time. Running alcohol through the injector pump will freeze it up if there is not sufficient oil to lubricate the pump. If it were not for the difficulty in getting the oil and alcohol to mix the diesel engine would be best.

You could put a carburetor on the air intake(if you can get to it) and mix the fuel about half and half, diesel through the injector and alcohol through the air intake. Again, watch out. The horsepower will go up roughly 10% to 20%.

PROPANE ENGINES

Alcohol can be used in propane engines if it is vaporized before it enters the carburetor. This vaporization is quite uncomplicated with alcohol. The alcohol you are using will vaporize at the temperature where it condensed in the still. These temperatures can be achieved by running the fuel intake through some of the cooling water as it leaves the engine. A small electric heater can be added to vaporize the fuel needed to start the engine and warm the water at first.

ALCOHOL POWERED CARS

The most efficient alcohol engine would vaporize the fuel before it entered a vapor carburetor, have a spark ignition system and compression ratio of 15 to 1. The engine would theoretically get about 100 miles to the gallon.

There are alcohol powered cars available in other countries. Brazil has been promoting the use of alcohol powered cars for many years now. Several companies and individuals have tried importing these Brazilian alcohol vehicles in recent years. They have not been tested by United States agencies to meet safety tests, but they will certainly meet emissions requirements. If you choose to try to import a Brazilian alcohol car you will need to sign a statement for the Department of Transportation that it is for experimental purposes and that you are aware they do not meet D.O.T standards.

CHARACTERISTICS OF FUELS

Table VIII-1 shows a variety of properties of gasoline, ethanol and #1 diesel fuel. These properties affect the performance of the fuel under various conditions.

The heating values are a measure of how much energy can be released when a fuel is burned, as in a heater. Notice that ethanol has a much lower heating value than the other two fuels. This means that in a heater it will produce less heat. In a standard heater you get 70% burning efficiency. Thirty percent of the fuel goes up in smoke. Alcohol can burn with 100% efficiency. This means that you can get about 75,670 BTUs from a gallon of alcohol and 98,000 BTUs from #2 fuel oil. The fuel oil will need a chimney to get rid of the 30% that becomes smoke while the alcohol heater can be vented directly into the room if simple provisions are made for the carbon dioxide. A chimney without a damper can take as much as 90% of the heat produced by a fire. Proper damper installation and an efficient furnace can reduce this to about 20%. This means you could possibly get 78,400 BTUs into your house from #2 fuel oil and still only 75,560 BTUs from alcohol. Still, with the alcohol you can be humidifying at the same time and be comfortable at a lower temperature, and there is no odor. If you make the fuel yourself at very low cost, provided you use the byproducts, you have the satisfaction of self-sufficiency.

There are several numbers in the table that have to do with the relative safety of the fuel. First, the flamability limits. There must be a whole lot more alcohol vapor in the air than gasoline vapor for it to explode. Therefore it would take a much larger leak in an alcohol system to cause an explosion than in a gasoline system. We must also use about 40% more alcohol in the engine to get complete combustion. That is why the carburetor should be changed in automobiles. Gasoline has a definite smell and it is easy to detect a gasoline leak. Pure alcohol has no odor, but there is enough water, carrying the odor of the mash, in homemade alcohol to allow the detection of alcohol leaks.

The autoignition temperature shows how hot the fuel must get before it will start to burn. Alcohol will not start burning until it is at a much higher temperature than gasoline. An added danger is that if you do get an alcohol fire started the flame is invisible in daylight. The easiest way to put out an alcohol fire is with water. The water mixes with the alcohol, diluting it to the point it can no longer burn. Water is necessary in the production of alcohol and therefore should always be available in the area.. A CO_2 fire extinguisher can also put out an alcohol fire and can be kept full with the CO_2 production of the alcohol and carried in any vehicle that runs on alcohol. If you even suspect that an alcohol fire may have gotten started, immediately douse the area with water or CO_2. If you see smoke or flames the alcohol has started something else on fire.

Never try to put out a gasoline fire with water. The gasoline will float on top of the water and continue burning, spreading the fire.

The flash point gives us some interesting information. The flash point is the temperature at which a spark will start an explosion in the proper mixture of air and vapor. Alcohol will not explode until it reaches 70°F. This causes some problems when starting the engine but is very convenient in the storage of fuel.

If the temperature of the fuel is below 70°F. there is no danger of the fuel exploding even if there is a leak in the storage tank. Cigarettes, light bulbs and other heat sources can heat the fuel above 70°F. so do not smoke, weld or allow any other spark in the immediate vicinity of an operating still. Store the fuel outside in a shaded or underground tank. If you treat it as you would gasoline you should have no safety problems.

The boiling point is another factor to consider. Some of the components of gasoline will start to boil at 80°F. That means that on warm summer days gasoline is boiling right out of any vented container. Alcohol does not boil until 173°F. Some of each liquid will evaporate at any temperature. The higher the vapor pressure the more liquid will escape through the vent. At temperatures below 80°F more alcohol will evaporate than gasoline, but not enough to be concerned about if the fuel is stored in a closed container.

TABLE VIII-1. SELECTED LIQUID FUEL CHARACTERISTICS

Thermal Properties	Gasoline	Ethanol	#1 Diesel
Lower Heating Value			
BTU/lb.	18,900 (avg.)	11,500	18,250
BTU/gal	115,400 (avg.)	75,670	133,332
Higher Heating Value			
BTU/lb @ 68°F.	20,260	12,800	19,250
BTU/gal	124,800		
Heat of Vaporization			
BTU/lb.	150	396	115
BTU/gal	900	3,378	
Octane Ratings			
Research	91-105	106-108	
Pump(RON+MON)/2	87- 98	98-100	10-30
Cetane #	10-20	-20-8	45
Flamability Limits	1.4-7.6	4.3-19.0	
(% Volume in Air)			
Stoich, A/F Ratio by weight	14.7:1	9:1	
BTU/ft^3 @ Stoich. A/F, 60°F.			
1 Atm. Gaseous Reactants			
BTU/ft.^{3}	95	93.8	
BTU/lb	210		
Specific Heat			
BTU/lb-°F.	0.48	0.60	
Autoignition Temp.	430-500	685	
Flash Point °F.	-50	70	
Coefficient of Thermal Expansion			
@ 60 °F. , 1Atm.	0.6×10^{-3}	1.12×10^{-3}	
Specific Gravity	.070-.078	0.794	.876
Liquid Density	~43.6	49.3	
Vapor Pressure			
Psi @ 100 °F.	7-15	2.5	
psi @ 77 °F.	~0.3	0.85	
Boiling Point °F.	80-440	173	
Freezing Point °F.	~-70	-173	
Viscosity @ 65 °F. (cp)	0.288	1.17	

The table also shows why many theorists, working with figures only, will tell you that you cannot get the same amount of work out of alcohol as you can out of gasoline. The following formulas show the relationship among work, energy, force and power.

Energy = Heat + Work   Work = Force x Distance   Force = Mass x Acceleration

Power = work/time or ½ V^2/time

In the internal combustion engine we want to do the work of moving the mass of the piston with as much acceleration as possible. The thing that does this work is the expansion of gasses following the explosion of the fuel. As you can see from the BTUs/ft^3 @ Stoich., A/F (that means the energy of a perfect explosion) about the same amount of heat is released in a gasoline explosion and an alcohol explosion. However, we used 1.4 times the alcohol.

The specific heat of a substance is how many BTUs it will take to raise one pound of a substance one degree. After combustion there is a mixture of gasses in each case. Because the alcohol mixture has more water in it - a high specific heat substance - the temperature will not rise as much in the alcohol mixture as in the gasoline mixture.

All gasses expand at approximately the same rate with a given temperature rise. The gasoline, which produces a higher temperature rise, is theoretically the better fuel. So more work can be done by gasoline if all the BTUs go into doing work.

Under less than ideal conditions the BTUs go into heat energy, not work. Engines do not create ideal conditions for work to be done. There is the octane rating to consider. Octane is a rating of how much the fuel can be compressed before it explodes. The more compression the more force the explosion will produce so more work can be done. Alcohol's higher octane rating allows us to compress it more and allows the piston to move farther after the explosion, making a more efficient engine. More of the BTUs of the explosion go into doing work and less into making the engine hot. Generally (and this depends on the carburetion and several other factors as well) the higher the compression ratio the more efficient the engine. A gasoline engine run on alcohol will use more fuel than it would have on gasoline. But an engine designed and built for alcohol can go farther on a gallon of fuel that an engine built for and run on gasoline.

CHAPTER 9

DISTILLER'S GRAIN USE

The farm alcohol producer is able to do something that the commercial distiller would never consider. That is overloading the brew with grain. Let's see what happens when you do that. The fermentation process takes out the carbohydrates, or energy producing components out of the grain. Distilleries have to buy all of the grain and must get all of the alcohol out of the bushel of grain that is economically feasible. On the farm the grain was grown to feed livestock. If there is still some carbohydrate left in the grain we haven't taken out all the energy.

By using 30 gallons of water per bushel of grain all, or nearly all, of the carbohydrates can be fermented. But if we only use 15 gallons of water the limiting factor becomes how long the yeast can live. Yeast will kill itself out in 12% to 15% alcohol. The yeast will make alcohol until that percentage of alcohol is in the brew, then will die. This will leave some of the energy in the grain untouched. The feed grains now provide the animal not only with protein but with some of the energy it needs to live on. We have already shown that the higher the alcohol percentage in the brew the easier it is to get out. With this method we have less alcohol per bushel of grain but it is easier to get out. Table VIII-3 gives the nutritional analysis two samples from which slightly less than two gallons per bushel of alcohol was produced.

TABLE VIII - 1 The amount of a substrate left after complete fermentation.

Material	Unit	% Fermentable	Lbs. Used	Lbs. Unused
Barley	Bushel	54.3	26.0	22.0
Buckwheat	Bushel	57.2	27.5	20.5
Corn	Bushel	57.8	32.4	23.6
Malt, Dry	Bushel	60.6	20.6	13.4
Oats	Bushel	43.6	14.0	18.0
Rice, rough	Bushel	54.6	24.6	20.4
Rye	Bushel	54.0	30.2	25.8
Wheat	Bushel	58.6	35.2	24.8
Corn Sugar	Bag	100	100	0.0
Molasses				
Blakstrap	Gallon	51.0	6.0	5.75
High test	Gallon	72.0	8.46	3.29
Sorgo Syrup	Gallon	51.0	6.0	5.75
Sugar Beets	Ton	16.0	320.0	1680.0
Sugar Cane	Ton	11.0	220.0	1780.0
Sucrose	Bag	100.0	100.0	0.0
Jerusalem Artichokes	Bushel	15.2	9.4	50.9
Carrots	Bushel	7.5	4.1	51.9
Potatoes	Bushel	15.6	9.4	50.6
Sweet Potatoes	Bushel	23.3	12.4	42.2
Yams	Bushel	18.7	10.3	44.7

The protein content of the feed depends on how much protein was in the material to start with and how much carbohydrate is left in the grain. The more water you add, within the limits described earlier, the less carbohydrates are left in the grain - provided you followed the cooking and liquifaction process correctly - and the more concentrated the protein. Only a small amount of protein has been added to the grain in the form of yeast organisms. This small amount of protein contains more Lysine than any other essential amino acid. Lysine is the amino acid that is limited in grain and the amount of Lysine in yeast helps balance the protein in the grain.

TABLE VIII-2 The nutritional analysis of 16 g. (½ oz.) Brewer's yeast

AMINO ACIDS

Isoleucine	456 mg.	Arginine	376 mg.
Leucine	504 mg.	Aspartic Acid	656 mg.
Lysine	584 mg.	Cystine	72 mg.
Methionine	96 mg.	Glutamic Acid	1224 mg.
Phenylalanine	352 mg.	Glycine	296 mg.
Threonine	384 mg.	Histidine	120 mg.
Tryptophan	88 mg.	Proline	320 mg.
Valine	416 mg.	Serine	312 mg.
Alanine	440 mg.	Tyrosine	272 mg.

MINERALS

Calcium (carbonate)	240 mg.	Potassium	320 mg.
Phosphorus	190 mg.	Zinc	.9 mg.
Iron	1.2 mg.	Manganese	.1 mg.
Copper	.37 mg.	Sodium	30 mg.
Manganese (oxide)	150 mg.		

Table VIII-2 gives the nutritional analysis of 16 grams of yeast. Because growing conditions vary so greatly there is no rule for determining the amount of yeast in the ration. The concentration of protein in the distillers grain is about three to four times the concentration in the original grain if all the carbohydrates have been removed. In the incompletely fermented samples tested it is a little less than twice as much.

Table VIII-3 is an analysis of two samples of farm produced distiller's grain compared to the analysis of the cracked grain before the fermentation process. Sample #3 had 64.72% moisture. It was strained through a window screen. The other distillers grain sample, # 1, was hand squeezed and allowed to air dry for a short time, during which time the yeast was growing on the grain. The yeast content, we believe, is what accounts for the high calcium content. The difference in protein content between the two samples is only .3%, not a significant difference

TABLE VIII-3 Analysis of farm produced distiller's grain after the production of approximately 2 gallons alcohol per bushel.

DRY MATTER BASIS

	Squeezed Distiller's Grain #1	Cracked Milo #2	Very Wet Distillers Grain #3
Protein	19.94	11.19	20.24
Phosphorus	.239		.255
Calcium	1.762		.215
Fiber	3.77	1.75	3.09
T.D.N.	88.83	89.55	89.99

.

Less than two gallons of alcohol per bushel was removed from these samples, meaning that some of the carbohydrates were still present. If liquifaction and fermentation had been complete the protein percentage would have been higher. Remember that the final protein content of the distillers grain is entirely dependant on the amount of protein in the grain you start with and the thoroughness of the fermentation process.

More important than how much protein that is in the rations fed to animals is the amount of that protein that is utilized by the animals. The digestive process is already started with the distillers grain and protein utilization is more complete. In ruminants the distillers grain bypasses the rumen and is digested more completely with less energy going into the digestion process. The yeast also contributes to the health of the digestive system of the animal and increases the efficiency of the digestion of roughage, according to reports by some veterinarians and researchers.

Distiller grain can make up as much as 17% of the total ration and it would still be an appetizing ration. This will provide about half the protein needs of the animal. This limit is determined by what the animal is willing to eat, not by the nutrient value of the feed grain. Combining the distillers grain with urea increases the utilization of urea and could significantly reduce the cost of protein feed rations.

Going back to Table I-1 we see that the first six items are all grains. All of these would have good feed value after processing. Table VIII-2 shows the percent fermentable of several possible substrates and the pounds used and unused. With grains the number of pounds left from a bushel is not great. For example with wheat we see that putting one bushel or 60 pounds in we would take out 25 pounds of unused grain and 2.56 gallons of 200 proof alcohol. Compare this to sugar cane, where one ton (2000 pounds) is put in and 1780 pounds must be removed. With silage of cane sorghum you would have even more fibrous material. However with silage there is some grain. If the silage were ground fine enough to crack the grain you would be making alcohol from the starch in the grain as well as the sugar in the stalks.

When tubers and roots are fermented the end product is a smelly mess with very little feed value. It is probably best used as fertilizer and plowed into the field as soon as possible. Fruits and vegetables will also smell if the pulp is left in the fermentation vat but can be used as feed if the juice is squeezed out and the pulp used while it is fresh. It does not have the same protein value as distillers grain. Everything is used when sugar is fermented and there is no feed product.

COMPLIMENTARY PROTEINS

Proteins are made up of building blocks called amino acids. There are twenty-two different amino acids that are necessary for animal life. Most of these amino acids can be made by the body but there are some that cannot be made by the bodies of animals. These are the essential amino acids. The amino acids that are essential vary somewhat from animal to animal, although there are a few amino acids that are essential amino acids to all animals. The essential amino acids for humans are tryptophan, leucine, isoleucine, lysine, valine, threonine, the sulfur containing amino acids and the aromatic amino acids. These amino acids are also essential to hogs.

Figure VIII-1 Relative necessary proportions of essential amino acids and portions of amino acids used when lysine is lacking.

In order for the body to function properly all of the essential amino acids must be available in the proper proportions. If any of the essential amino acids are lacking none of them can be used completely. For example, if adequate amounts of all the essential amino acids were present except for lysine and only 50% of the necessary lysine was present only 50% of all the essential amino acids would be used. See Figure VIII-1

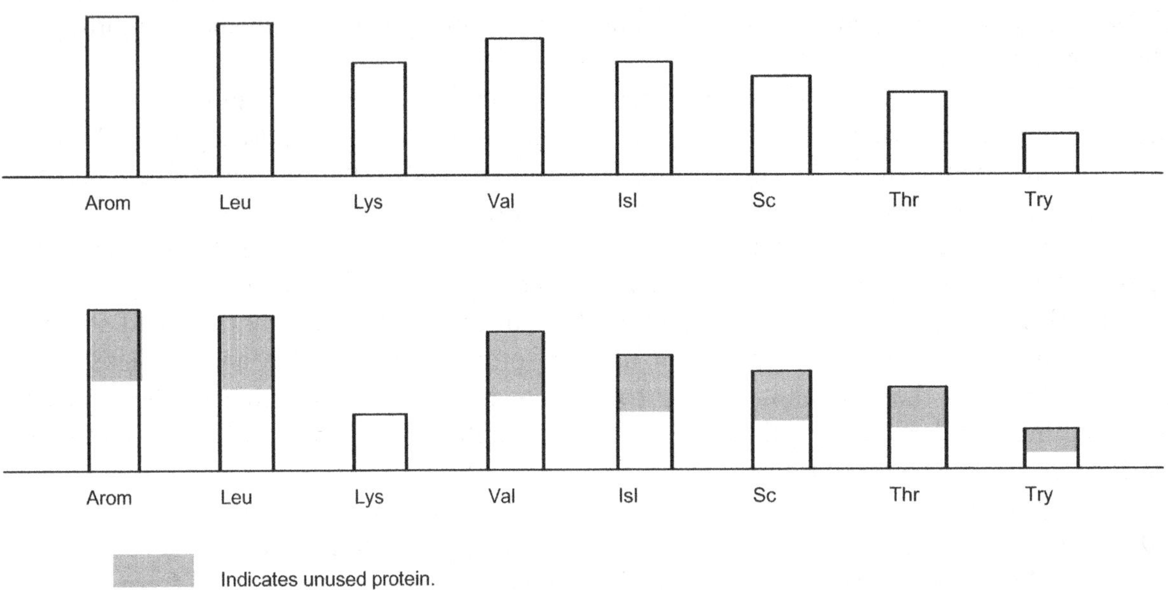

In order to have a balanced ration the amino acids must be balanced. Figure VIII-2 shows the relative amounts of the most commonly inadequate amino acids in some substrates. These figures can be used to make a more balanced ration. For example, soybeans are high in lysine and isoleucine but lacking the sulfur containing acids. Wheat and corn lack lysine and isoleucine but are high in the sulfur containing amino acids. By mixing soybeans and wheat you can make a more complete supplement for animals.

Figure VIII-2 Amino acids provided by various grains.

VALUE OF DISTILLERS GRAIN

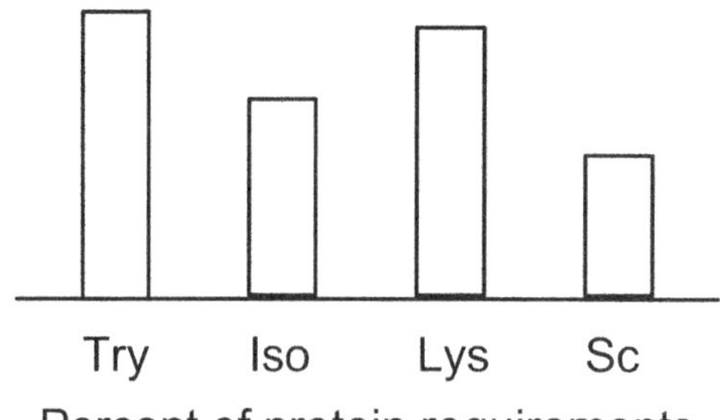

Percent of protein requirements
supplied by soybeans.

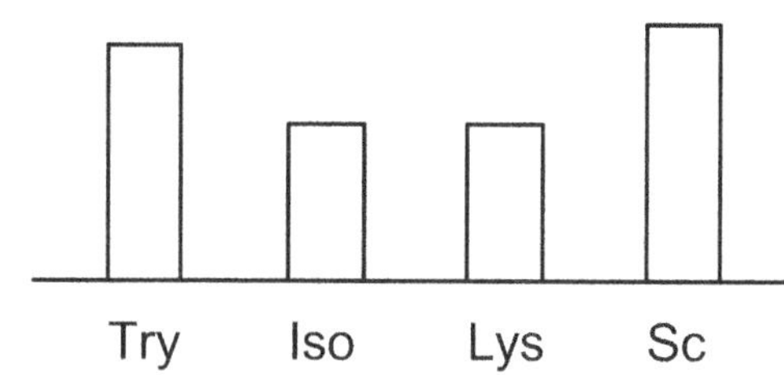

Percent of protein requirements
supplied by buckwheat, oats, and barley.

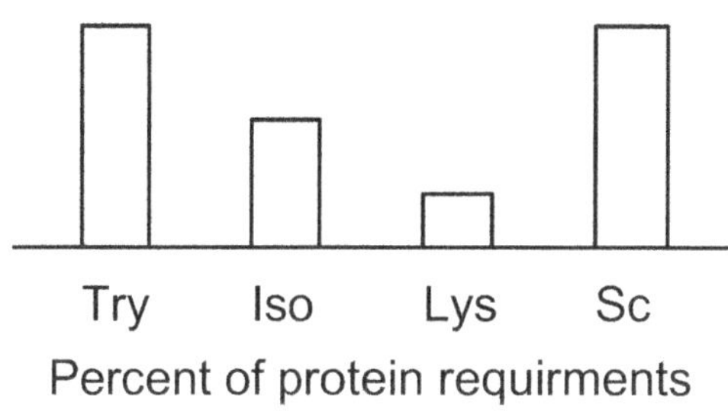

Percent of protein requirments
supplied by wheat.

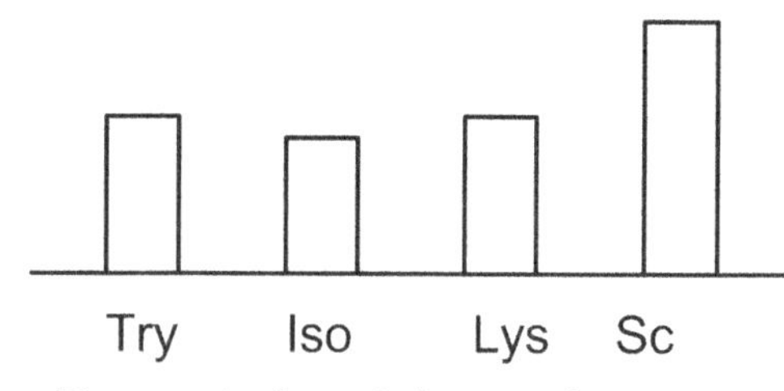

Percent of protein requirements
supplies by corn and rye.

Most of the research that has been done on distiller's grain has used dry distiller's grain. There is some research on wet distiller's grain. This research shows the wet distillers grain had widely varying values in different uses.

Research at the University of Nebraska shows that calves to 500 pounds gain weight twice as fast on distillers grain as on a protein source such as soybean meal, provided they consume equal amounts of protein. Because the soybean meal is 40% protein and the distillers grain tested 25-30% protein they cannot be compared directly pound for pound. Using a University of Nebraska formula it would take 1,853 pounds of distillers grain (dry weight), 135 pounds of urea and 12 pounds of corn to equal a ton of soybean meal for calves. At the time of this test the ton of soybean meal cost $210. Subtracting the cost of the urea ($10.80) and the cost of the corn ($.60) gives a value of $198.60 for the 1853 pounds of distillers grain or $.11 per pound. If 23.6 pounds of distillers grain were produced from a bushel of corn the value of the distillers grain per bushel would be $2.60 at 1979 prices. If the solubles (that part of the grain dissolved in the water) were included in calf rations the efficiency of the utilization of the protein dropped so that it took 2,346 pounds of distillers grain plus solubles (dry weight) and 87 pounds of urea to equal a ton of soybean meal. Larger feedlot animals show reduced utilization of the distillers grain and a larger quantity should be made available for feedlot cattle.

The solubles are utilized more efficiently by hogs and chickens. There are problems associated with the amount of water in the feed. If you are feeding hogs or poultry the following procedure will work for increasing the value of the feed. After the second day of fermentation open the top of the tanks and stir until you are ready to distill. This will reduce the alcohol yield by 10% or so but will encourage reproduction of yeast by introducing oxygen. If you do this be sure the temperature remains above 90° to reduce the likelihood of contamination.

Follow the normal distillation procedure then drain the boiler into another tank. Allow it to settle as it cools. When you start the next batch use the cool water, taking the water from the top of the settling tank, then drain the sludge away from the bottom troughs for hogs or chickens. The main value of the solubles as food for hogs and poultry lies in the fact that they are high in phosphorus and B vitamins. The solubles can be as much as 1.5% phosphorus in the form most available to hogs and poultry. Only 7% of the phosphorus is in phytate form, a form hogs and poultry cannot use well.

Feeding distiller grains plus solubles to gilts at a rate of 44.2% of the total diet showed no effect on the number of live pigs at birth, the number of live pigs at weaning or weaning weight. It did result in reduced weight gain by the sows during gestation.

Feeding the solubles to young pigs increased their weight gain over feeding soybean meal as a protein source. Three trials ranged from a 6% greater weight to a 34% greater weight. Weight gains on a ration of distillers grain plus solubles were above gains using soybean meal but below those using just the solubles as a protein source.

Because the distiller's grains have a higher fat and fiber content than soybean meal and many other protein sources they are especially valuable to dairy cattle. When the distillers grain plus solubles are fed butterfat has been shown to increase .2% to .3%. Milk production increased .5 pound over soybean meal, 1.7 pounds over hominy feed and 2.9 pounds over cottonseed meal per animal per day. The distillers grain can replace around 40% of the grain ration.

When feeding either wet distillers grain or the solubles there is a chance of mold. Molds could cause problems because they are unappetizing and could cause digestive problems or even poisoning. Uneaten solubles should be discarded each day. If the distiller's grains are removed before distillation the uneaten portion should be removed every one to three days depending on the weather. If you boiled the distiller's grain during distillation then removed and strained them while hot you will have dry or almost dry distiller's grain and should remove the uneaten portion only when it appears necessary. Where removal would be a problem the expense of removing the spoiled feed should be considered and balanced against the increased income resulting from the feed.

Most owners of on the farm stills seem to feed the mash as it comes out of the tank by simply emptying into a tank or trough and letting the livestock eat it at will. However many are looking for a more scientific way of making sure their animals get the nutrients they need. The rations on the following pages were formulated through the least cost feedmix program on Nebraska's Agnet computer. The nutrient requirements of different animals and nutrient analysis of different feedstuffs were prepared by the University of Nebraska.

Because we were looking specifically for rations using distillers' grains and because the computer would always recommend the least expensive ingredients we assigned a low cost to the distillers grain and a high cost to soybean meal. In all but the dairy rations, if corn costs $3.05 a bushel or more and the entire cost of the corn is assigned to the distillers grain (resulting in very inexpensive alcohol) the economics of the rations that follow would change. If the whole picture is considered, the cost to the farmer for feed and fuel combined would be less than feeding conventional rations even though in some cases half the cost of the grain must be assigned to the alcohol for the feed to be less expensive. In the dairy ration the mash could cost as much as $195 a ton without changing the ration, a pretty high cost for anything that is 70% water.

In the swine ration lysine and tryptophan are at least 10% above minimum requirements in all cases and all other nutrients meet minimum requirements. Fat and fiber are well below minimum levels.

For range cattle the crude protein for beef cattle is almost three times the minimum requirement for cattle on fertile pastures so this supplement can produce healthy animals on very poor pasture. In most cases the crude protein for beef cattle is almost twice the minimum requirement although the ration is still balanced. The same is true for horses.

For poultry the mash can be fed to pullets and laying hens as it comes from the tank, although you may wish to screen it first and feed them only the grain as they are more likely to eat it that way.

SWINE

Pre-starter, 22% protein.   Early weaned or orphan pigs

Feed Name	% Moisture	Ration % as Fed	Ration 90% Dry
Distiller grain (corn)	70.00	15.24	5.88
Corn	14.00	50.20	54.78
Soybean meal	11.00	31.30	35.36
Limestone	0.00	0.67	0.85
Phos. Dical	4.00	1.54	1.87
Salt	0.00	0.22	0.28
Tm. Premix	0.00	0.04	0.06
Vit. Premix	0.00	0.79	1.00
Totals		100.00	100.00

Starter, 18% Protein. Two weeks of age to 40 lbs. body weight

Distiller grain (corn)	70.00	23.14	9.38
Corn	14.00	53.00	61.56
Soybean meal	11.00	20.69	24.88
Limestone	0.0	0.59	0.79
Phos. Dical	4.00	1.58	2.05
Salt	0.0	0.21	0.28
Tm. Premix	0.0	0.05	0.06
Vit. Premix	0.0	0.74	1.00
Totals		100.00	100.00

Starter, 18% Protein.   Two weeks of age to 40 lbs. body weight

Soybean meal	11.00	24.27	25.61
Limestone	0.0	0.72	0.85
Phos. Dical	4.00	1.72	1.95
Salt	0.0	0.23	0.28
Distillers grain (milo)	70.00	7.37	2.62
Milo	12.00	64.80	67.63
Tm. Premix	0.0	0.05	0.06
Vit. Premix	0.0	0.84	1.00
Totals		100.00	100.00

Grower Finisher, 16% protein.   40-100 lbs. body weight.

Distiller grain (corn)	70.00	32.49	14.23
Corn	14.00	52.68	66.15
Soybean meal	11.00	12.48	16.22
Limestone	0.0	0.75	1.09
Phos. Dical	4.00	0.69	0.97
Salt	0.0	0.19	0.28
Tm. Premix	0.0	0.04	0.06
Vit. Premix	0.0	0.68	1.00
Totals		100.00	100.00

Grower Finisher, 16% protein. 100-170 lbs. body weight

Soybean meal	11.00	8.56	10.64
Limestone	0.00	0.87	1.22
Phos. Dical	4.00	0.58	0.78
Salt	0.00	0.20	0.28
Distillers grain (milo)	70.00	28.93	12.12
Milo	12.00	60.10	73.90
Tm. Premix	0.00	0.04	0.06
Vit. Premix	0.00	0.72	1.00
Totals		100.00	100.00

Finisher, 10% Protein. 175 lbs. to market weight

Distiller grain (corn)	70.00	27.34	11.53
Corn	14.00	64.66	78.16
Soybean meal	11.00	5.86	7.33
Limestone	0.00	0.57	0.80
Phos. Dical	4.00	0.63	0.85
Salt	0.00	0.20	0.28
Tm. Premix	0.00	0.04	0.06
Vit. Premix	0.00	0.71	1.00
Totals		100.00	100.00

Gestation - Summer 3-3.5 lbs./head/day or 4500 Kcal/head/day

Distiller grain (corn)	70.00	25.50	10.53
Corn	14.00	56.50	66.89
Soybean meal	11.00	13.36	16.37
Limestone	0.00	0.48	0.66
Phos. Dical	4.00	3.19	4.22
Salt	0.00	0.20	0.28
Tm. Premix	0.00	0.04	0.06
Vit. Premix	0.00	0.73	1.00
Totals		100.00	100.00

Gestation - Winter 6-6.5 lbs./head/day or 9000 Kcal/head/day

Distiller grain (corn)	70.00	31.31	13.65
Corn	14.00	64.23	80.25
Soybean meal	11.00	2.06	2.66
Phos. Dical	4.00	0.94	1.31
Salt	0.0	0.19	0.28
Tm. Premix	0.0	0.04	0.06
Vit. Premix	0.0	0.69	1.00
Totals		100.00	100.00

POULTRY

Pullet Developer 16% protein

Distiller grain (corn)	70.00	22.04	8.78
Corn	14.00	56.33	64.29
Limestone	0.00	0.16	0.21
Phos. Dical	4.00	0.88	1.13
Salt	0.00	0.32	0.42
Tm. Premix	0.00	0.03	0.04
Lyamine	0.00	0.07	0.09
MHA	0.00	0.02	0.02
Vit. Premix	0.00	0.31	0.42
Alf. Dehy. 17	7.00	10.83	13.36
Poultry BPM	6.00	9.01	11.24
Totals		100.00	100.00

Pullet Grower 17% Protein

Distiller grain (corn)	70.00	232.48	9.44
Corn	14.00	54.33	62.64
Phos. Dical	4.00	0.91	1.17
Salt	0.0	0.30	0.40
Tm. Premix	0.0	0.03	0.04
Lyamine	0.0	0.07	0.10
MHA	0.0	0.05	0.07
Vit. Premix	0.0	0.31	0.42
Alf. Dehy. 17	7.00	10.00	12.47
Poultry BPM	6.00	10.51	13.25
Totals		100.00	100.00

BEEF

Prairie Hay, 400-500 lb. Steers and heifers. 1.5-2 lb. ADG

Distiller grain (corn)	70.00	80.89	58.51
Corn	15.50	0.36	0.74
Prairie Hay	10.00	18.43	40.00
Limestone	0.00	0.18	0.44
Phos. Dical	4.00	0.13	0.31
Totals		100.00	100.00

Prairie Hay Finisher (7-14 days)

Distiller grain (corn)	70.00	37.17	16.86
Corn	15.50	31.66	40.45
Prairie Hay	10.00	18.43	40.00
Animal Fat	0.00	1.32	2.00
Limestone	0.00	0.27	0.41
Phos. Dical	4.00	0.18	0.28
Totals		100.00	100.00

Alfalfa Hay, 300 lb. Steers and heifers, 1.5-2 lbs. ADG

Distiller grain (corn)	70.00	79.19	56.02
Corn	15.50	1.79	3.56
Alfalfa Hay	10.00	18.85	40.00
Phos. Dical	4.00	0.18	0.42
Totals		100.00	100.00

Alfalfa Hay, 500-700 lb. Steers and heifers, 1.5-2 lbs. ADG

Distiller grain (corn)	70.00	71.64	45.71
Alfalfa Hay	10.00	28.36	54.29
Totals		100.00	100.00

Finisher 2 (7-14 days)

Distiller grain (corn)	70.00	57.24	31.43
Corn	15.50	18.47	28.57
Alfalfa Hay	10.00	24.28	40.00
Totals		100.00	100.00

Bred 1-2 years .5 lb. Gain (+ fetal gain)

Distillers grain	70.00	99.91	99.71
Limestone	0.00	0.09	0.29
Totals		100.00	100.00

Bull 1700 lbs., no gain

Distillers grain	70.00	99.92	99.74
Limestone	0.00	0.08	0.26
Totals		100.00	100.00

Range, 500 lb. Calves, 1.5 lb. ADG

Feed Name	% Moisture	Ration(%) as fed	LB/HD/DAY Dry	As Fed
Distillers Grain	70.00	75.00	6.40	21.33
Meadow Fer.	10.00	11.88	3.04	3.38
Grower 40	10.00	10.21	2.61	2.90
Cubes 13%	10.00	2.87	0.73	0.82
Salt	0.00	0.05	0.01	0.01
Totals		100.00	12.80	28.44

Cows nursing calves, Average milking cow

Distillers Grain	70.00	78.32	10.26	34.19
Meadow, Dry	10.00	21.54	8.46	9.40
Salt	0.00	0.14	0.06	0.06
Totals		100.00	18.78	43.65

DAIRY

High Producers, Roughage and grain mix (regular), Fed separately.

Feed Name	% Moisture	Ration(%) as fed	LB/HD/DAY	
			Dry	As Fed
Distillers Grain	70.00	37.05	4.80	19.58
Prairie Hay	10.00	16.78	7.98	8.87
Soybean Meal	11.00		6.28	2.95
				3.32
Limestone	0.00	0.57	0.30	0.30
Phos. Dical	0.00	0.21	0.11	0.11
Tm. Salt	0.00	0.26	0.12	0.13
Vitamin A	0.00	0.05	0.03	0.03
Corn	15.50	24.90	11.12	13.16
Sil. Corn	65.00	13.90	2.57	7.34
Totals		100.00	30.00	52.85

Medium Producers, Roughage and grain mix (regular), Fed separately.

Feed Name	% Moisture	Ration(%) as fed	LB/HD/DAY	
			Dry	As Fed
Distillers Grain	70.00	45.52	4.00	20.00
Prairie Hay	10.00	25.36	10.03	11.14
Soybean Meal	11.00		5.09	1.99
				2.24
Limestone	0.00	0.42	0.18	0.18
Phos. Dical	0.00	0.09	0.04	0.04
Tm. Salt	0.00	0.26	0.11	0.11
Vitamin A	0.00	0.07	0.03	0.03
Corn	15.50	23.19	8.61	10.19
Totals		100.00	25.00	43.94

Low Producers, Roughage and grain mix (regular), Fed separately.

Feed Name	% Moisture	Ration(%) as fed	LB/HD/DAY	
			Dry	As Fed
Distillers Grain	70.00	45.68	3.20	16.00
Prairie Hay	10.00	32.52	10.25	11.39

Soybean Meal	11.00		3.65	1.14
				1.28
Limestone	0.00	0.31	0.11	0.11
Phos. Dical	0.00	0.04	0.01	0.01
Tm. Salt	0.00	0.26	0.09	0.09
Vitamin A	0.00	0.05	0.02	0.02
Corn	15.50	17.49	5.18	6.13
Totals		100.00	20.00	35.03

Grain ration only (14% Protein) Feed with average legume forage.

Molasses Cane	25.00	5.88	1.20	1.60
Soybean Meal	11.00		1.49	0.36
				0.41
Limestone	0.00	0.55	0.15	0.15
Phos. Dical.	0.00	1.79	0.49	0.49
Tm. Salt	0.00	0.55	0.15	0.15
Vitamin A	0.00	0.18	0.05	0.05
Vitamin D	0.00	0.02	0.01	0.01
Corn	15.50	35.52	8.16	9.66
Distillers Grain	70.00	53.85	4.39	14.64
Totals		100.00	15.00	27.19

Dry cows, in very good condition and under usual weather conditions

Distillers grain	80.00	27.42	1.55	7.74
Prairie Hay	10.00	72.26	18.36	20.40
Phos. Dical	0.00	0.08	0.02	0.02
Tm. Salt	0.00	0.18	0.05	0.05
Phos. Monos.	0.00	0.06	0.02	0.02
Totals		100.00	20.00	28.23

HORSE

Mature Horse at light work on pasture

Distillers Grain	70.00	59.50	5.29	17.64
Meadow Fer.	10.00	33.24	8.87	9.85
Grower 40	10.00	7.04	1.88	2.09

Pho. Sod. TP	4.00	0.02	0.01	0.01
Salt	0.00	0.20	0.06	0.06
Totals		100.00	16.11	29.65

DISTILLER'S GRAIN FOR CATFISH

Distiller grain solubles are almost universally used in fish rations for a variety of fish. Distiller's grain can also be used for up to 45% of the ration. Catfish need some animal protein, which is generally provided by fish meal. The ration below will provide all the nutritional needs of chanal catfish.

Feed Name	% Dry
Fish meal	20
Distillers Grain	35
Soybean Meal or Peanut Cake	15
Alfalfa Meal	5
Grain Byproducts (e.g. Rice Bran)	15
Rice Hull Fractions	8
Demineralized salt (not required in some water)	1
Vitamin Premix	1

The vitamin premix should contain the following

Vitamin A	450,000 USP units
Vitamin D_3	200,000 IC
Riboflavin	300 mg
Pantothenic Acid	600 mg
Niacin	3,500 mg.
Choline Chloride	40,000 mg
Vitamin B_{12}	1 mg
Vitamin E	150 IU
Vitamin K (Mendione Sodium Bisulfate)	100 mg.
Ethoxyquin	6.5 g
Folic Acid	40 mg

One problem with feeding the distillers grain directly from the tank is that the small particles of the feed reduce conversion efficiency by 50%, meaning you must feed twice as much to get a gain equal to feeding larger pellets. You also risk pollution if the fish do not eat all the grain, resulting in over fertilization of the pond and algae blooms. If you plan to drain your tanks directly into any fish pond an aeration system should be provided or raceway culture considered.

The following table shows the feeding rate for channel catfish at different temperatures.

Water Temperature °C.	Daily Weight Fed as a % of Total Weight of Stock
More than 32 °C.	Up to 1.5%
21-32 °C.	3%
16-21 °C.	2%
7-16 °C.	1%
Less than 7 °C.	0.5% on warm sunny days only

CHAPTER 10

PERMITS

In order to legally produce alcohol you must have a permit from the Bureau of Alcohol, Tobacco and Firearms. The Bureau has been extremely helpful with fuel alcohol producers and has issued a number of permits for fuel alcohol production. For information on the latest regulations contact www.atf.treas.gov or natlrevctr@cinc.atf.treas.gov. A direct line to find out about applications for permits for fuel alcohol production is 513-684-7150. You can receive an application form and instructions for filing it from these sources. The form and process are relatively simple. You will need to file a report annually on your production and how it was used or disposed of.

Permits are issued for small, medium and large plants. The requirements for each size plant are listed below.

A.   The Small Plant:

1. Definition: A plant producing or receiving no more than 10,000 proof gallons per year. (A proof gallon is 1 gallon of a liquid at 60°F. that contains 50% alcohol. A gallon of 180 proof alcohol would equal 1.8 proof gallons.

2. Permits will be approved within 60 days, provided the application is completed correctly. If it is not completed correctly it will be returned for correction within 15 days.

3. Sale or disposition of fuel alcohol (alcohol rendered unfit for beverage purposes) is not restricted.

4. The permit application is form # 5110.74, available from the Bureau of Alcohol, Tobacco and Firearms.

5. An annual report through December 31 will be required. The operator should keep records of alcohol production and disposition throughout the year.

B.   The Medium Plant

1. Definition: A plant producing and/or receiving 10,000 to 50,000 proof gallons in one year.

2. A bond is required. The face amount of the bond will be a minimum of $2000 to a maximum of $50,000, depending on the amount of production and receipts per year. When calculating the amount of the bond required figure the face amount of the bond at $1000 for each 10,000 proof gallons or fraction thereof.

3. Application for a permit uses the same form as the small plant with some additional information, particularly information involving the principal persons involved in the business and the organization of the business.

4. An annual report through December 31 will be required. The operator should keep records of alcohol production and disposition throughout the year.

5. Sale or disposition of fuel alcohol (alcohol rendered unfit for beverage purposes) is not restricted.

C.   <u>The Large Plant</u>

1. Definition: A plant producing and/or receiving more than 500,000 proof gallons per year.

2. A bond is required. The face amount of the bond will be $50,000 plus $2000 for each additional 10,000 proof gallons. The maximum bond is $200,000 regardless of yearly production or receipts.

3. Application for a permit uses the same form as the small plant with some additional information, particularly information involving the principal persons involved in the business and the organization of the business.

4. An annual report through December 31 will be required. The operator should keep records of alcohol production and disposition throughout the year.

5. Sale or disposition of fuel alcohol (alcohol rendered unfit for beverage purposes) is not restricted.

Bonds

The cost of a bond is similar to the premium paid for an insurance policy, that is, a fraction of the face amount. There are several types of bonds you may file. The bond may be executed on an ATF bond form by a corporate surety (a bonding or insurance company holding a certificate of authority from the Secretary of the Treasury.) In addition the bond must be accompanied by a power of attorney from the surety company authorizing the person who signs the bond for the surety company to act on the surety's behalf. This power of attorney is prepared on a form provided by the surety company and will need to be stamped with the corporate seal of the company. Your insurance broker should be able to handle this detail.

In lieu of a corporate surety you may pledge and deposit, as surety for your bond, securities which are transferable and guaranteed as to both interest and principle, by the United States. If you propose to submit a bond underwritten by securities you should contact the ATF office in your region for further instructions.

In place of either of the above cash may be accepted as security for your bond. If you propose to submit cash you should still complete an ATF bond form. The cash will need to be in the form of a postal money order, cashier's check or certified check made payable to the Internal Revenue Service. The bond form should be signed and dated in the presence of two witnesses who must also sign their names in the spaces provided.

GROUP FUEL PRODUCTION

If several people would like to get together to make alcohol for themselves and others there is a way to do this without the need to get a commercial permit. Simply form a corporation for fuel production. The person who is going to make the alcohol could be the head of the corporation. You could not sell the alcohol, even to each other, but it would belong to every member of the corporation. The producer(s) could be paid a wage for the work of producing the alcohol. All expenses would be handled by the corporation. Grain can be purchased from and feed grain sold to anyone, member or nonmember.

Before forming a corporation contact the ATF office in your region to get the current regulations. The ATF has been very cooperative with fuel alcohol production but only congress can make laws and the ATF must work within the framework of existing laws.

ATF Field Division Offices:

Bureau of Alcohol, Tobacco & Firearms
Baltimore Field Division
31 Hopkins Plaza
5th Floor
Baltimore, Maryland 21201
(410) 779-1700
Fax: (410) 779-1701

Bureau of Alcohol, Tobacco & Firearms
Boston Field Division
10 Causeway Street
Suite 253
Boston, Massachusetts 02222
(617) 565-7040

Bureau of Alcohol, Tobacco & Firearms
Columbus Field Division
37 West Broad Street, Suite 200
Columbus, Ohio 43215-4167
(614) 469-5303
Fax: (614) 469-5308

Bureau of Alcohol, Tobacco and Firearms
Chicago Field Division
300 South Riverside Plaza
Suite 350
Chicago, Illinois 60606
 (312) 353-6935
Fax: (312) 353-7668

Bureau of Alcohol, Tobacco & Firearms
Detroit Field Division
1155 Brewery Park Boulevard
Suite 300
Detroit, Michigan 48207-2602
 (313) 393-6000
Fax: (313) 393-6058

Bureau of Alcohol, Tobacco & Firearms
Kansas City Field Division
 2600 Grand Avenue
Suite 200
Kansas City, Missouri 64108
(816)421-3440
Fax:(816)421-6511

Bureau of Alcohol, Tobacco and Firearms
Los Angeles Field Division
350 South Figueroa Street
Suite 800
Los Angeles, California 90071
(213) 894-4812
Fax: (213) 894-0105

Department of the Treasury
Bureau of Alcohol, Tobacco and Firearms
Louisville Field Division
600 Dr. Martin Luther King Jr. Place
Suite 322
 Louisville, KY 40202
(502) 582-5211
Fax: (502) 582-5634

Department of the Treasury
Bureau of Alcohol, Tobacco and Firearms
Nashville Field Division
5300 Maryland Way, Suite 200
 Brentwood, Tennessee 37027
615-781-5364
Fax: 615-781-5371

Bureau of Alcohol, Tobacco & Firearms
111 Veterans Memorial Blvd.
 Suite 1008
 Heritage Plaza Building
 Metairie, Louisiana 70005
 (504) 841-7000
 Fax: (504) 841-7039

Department of the Treasury
Bureau of Alcohol, Tobacco and Firearms
 New York Field Division
6 World Trade Center
 Suite 600
New York, NY 10048
(212)466-5145
Fax: (212)466-5160

Bureau of Alcohol, Tobacco & Firearms
Philadelphia Field Division
United States Custom House
Suite 607
Philadelphia, Pennsylvania 19106
(215) 597-7266
Fax: (215) 597-6116

Bureau of Alcohol, Tobacco & Firearms
St. Paul Field Division
30 East Seventh Street
 Suite 1870
St. Paul, Minnesota 55101
 (651) 290-3092
Fax:(651) 290-3363

Bureau of Alcohol, Tobacco & Firearms
Tampa Field Division
501 East Polk Street
Suite 700
Tampa, Florida 33602
(813)228-2021
 Fax: (813)228-2111

Bureau of Alcohol, Tobacco & Firearms
Washington Field Division
607 14th Street, NW
Suite 620
Washington, DC 20005
(202)927-8810
Fax: (202)927-4024

Additional field offices can be found on the ATF web site with telephone number for contacting the most appropriate office in your region.

In the 1980s and early 1990s there were several tax benefits to using alcohol and several programs available for financing fuel production. Most of these programs have expired at this time but with the current rising price of crude oil there may be other credits and assistance programs available soon. Keep checking the Internet and newspapers for word of any new programs.

By now you can see that alcohol production is a complex activity. There are all kinds of things from which alcohol can be made. Some will produce a high quality feed product and others will not produce an edible product at all. A farmer must consider how many gallons per acre can be produced, not just bushels or tons to the acre, and also how much labor, in the fields and during the process, will be needed. There are some things that just won't work. Like green chopped alfalfa. It will produce tannic acid which will kill the yeast. Even some weeds can be fermented, but not if they produce high quantities of tannic acid. So we cannot say you can mix just anything and produce alcohol fuel. However, mixing grain with silage would produce good alcohol and a high quality feed product that would be high in roughage. You might want to set up a small experimental tank to text the results of mixing various substrates.

There are political and economic considerations that are mentioned throughout this book. You must not only make decisions on how you are going to make alcohol but on how to handle these other factors as well. At the time of this printing tax incentives for fuel alcohol have just expired. Hopefully this is a temporary situation. With the rapidly changing energy situation in the world it may soon be more beneficial to make alcohol than use fossil fuels even without the benefit of tax incentives.

Fuel alcohol is mainly ethanol, which is only slightly toxic, but it may have some more toxic contaminants in it. If anyone should drink any of your fuel alcohol proceed as follows:

1. CALL DOCTOR

2. INDUCE VOMITING

3. TAKE VICTIM TO HOSPITAL EMERGENCY ROOM

The main component of your fuel alcohol is high proof ethanol which can cause irritation of sensitive skin and is slightly toxic when taken orally.

Some possible contaminants, present as less than 1% of the total liquid are acetylaldehyde and isopropyl, isobutyl and isoamyl alcohol, all of which can cause severe skin irritation with prolonged contact and all of which are toxic.

While it would take about a quart of fuel alcohol to kill a 150 pound man it would take only a few ounces to kill a toddler so always store it in tightly capped containers out of the reach of children

Whenever any of the alcohol is ingested it will kill some brain cells and drinking less than lethal quantities can cause blindness and nervous disorders. Label all alcohol containers, whether the contents are denatured or not, as poison.

The vapors of ethanol and all of the contaminants can irritate the respiratory system if they are present in large concentrations so be sure to ventilate your still room if anything happens to cause the release of alcohol vapors into the room. Once you have opened windows and doors leave the room until the vapors have had a chance to dissipate.

You may want to post this page near your fuel alcohol storage area.

APPENDIX A

PLANS FOR A REMOTE POST COLUMN

These instructions will allow you to construct the Remote Post Column described in this book. This is a small still that can produce up to several gallons per hour of 180 proof alcohol. Before attempting to produce alcohol using this still read the entire book carefully and pay close attention to the safety warnings. Be sure you have a pressure relief valve on your boiler and be careful not to let alcohol vapors escape into your distillation room.

BOILER

You will need to provide a boiler as part of the equipment. Plans for a boiler are not included but this section includes several suggestions on how to provide the best boiler. The boiler should provide the largest liquid surface area possible. Vapor leaves the liquid only at the surface, so the more surface area the more opportunity for vapor to leave. There is also more area on which to provide heat, thus speeding up the distillation process. Consider, for example, a 30-gallon oil drum. If heat is applied to the end of the drum while it stands vertically the vaporizing surface would be the diameter of the drum. If it were lying on its side the heated surface and the vaporizing surface would be considerably greater. The boiler should be vapor and liquid tight and able to withstand heat and pressure.

The distilling process begins at the boiler and the plumbing from the boiler to the still becomes the first stage of the distillation process. If the piping from the boiler is not insulated much of the water will condense and run back into the boiler or out through the runoff outlet at the bottom of the column. A slight downward slant from the boiler to the column will prevent the water from draining back into the boiler. This simple plumbing trick will extend the capacity of the still greatly.

REMEMBER - All the still is doing is cleaning the water out of the vapors before condensing them. The higher the proof we want the more cleaning we want. As soon as the vapors come out of the boiler any slight cooling will strip a portion of the vapor out.

Follow the drawings for suggested proportions of the sections described below.

DOUBLER

The doubler is the bottom part of the still and where the vapors first enter it. Here we create a pool of liquid from the stripping above and make the incoming vapors pass through it. This does several things.

1.	The vapor heats the pool. In heating the pool the vapor will be cooled and a considerable amount of water will condense. The reason it is called a doubler is because just about half the water condenses out every time the vapor passes through one of these pools.

2.	The heated pool will be above the boiling point of alcohol so that any alcohol falling from above will be reheated and go back up.

3.	The alcohol is more concentrated at the top of the pool so the water is slipped out from under the alcohol through the runoff line.

4.   This pool will keep any solids that might have reached this point in suspension so that they will float out in the runoff.

This is a very important part of the still even though it isn't difficult to build. In order to keep the pressure that is built up as the vapor passes through the marbles from forcing the pool out the pool should be at least 7 inches deep. The depth of the pool is set by the trap made by the runoff line.

The first vapors to enter the column will encounter a cold pool which will condense some of the alcohol, lowering the boiling point of the pool. This alcohol will boil out later as the temperature of the incoming vapor rises because the boiler has less alcohol in it.

MARBLE STRIPPER

This is the second type of still that is used in most small operations. The Remote Post Still combines the two types.

The material in the stripper does not need to be marbles. It can be anything that will allow a great deal of surface area on which the water can condense, will not absorb or dissolve in water or alcohol and will not pack so tightly that the vapor will not be able to flow up and the water flow down. Nuts and bolts, brass or stainless steel wool (do not use regular steel wool), or even rocks will work. There is a plate at the bottom and one at the top of this section with many holes in them. These holes should be at least 3/16 inches in diameter so the liquid going down will not close the holes and stop the vapors from rising. They should be small enough that the stripper material does not fall through. The plate at the bottom supports the stripper material and the one at the top distributes the downward flow more evenly.

STRIPPER COIL SECTION

Now the vapors encounter the coil of tubing that will cool them slightly. This cooling will condense some of the water and some of the alcohol will follow it. As this condensate falls down it will tend to cool the marble stripper.

Once there are hot vapors coming up and cooled condensate coming down the vapors will be cooled by the condensate and condense on the stripping material. The falling condensate will be heated by upcoming vapors farther down in the still Because the water vapor will condense more readily than the alcohol vapors each condensation and vaporization increases the concentration of the alcohol in the vapor and water in the liquid. By adjusting the flow of water in the stripper coil we can attain just the right up and down flow to get the proof we want.

CONTROL SECTION

This section simply puts a distance between the top of the stripper coil and the condenser to keep the two from affecting each other and allows us to get some idea what is happening in the still. We need some reference temperature so we can set the proof of the alcohol coming from the still.

When the up and down flow of vapors and condensate in the stripper is just right the vapors that finally get through will be cleaned to the proof we want. A thermometer placed in the column a little above the middle of the control section will give us a reference temperature. Alcohol boils at sea level at 173° F. We don't want the temperature at our reference point below this or we will be condensing too much alcohol. In order to distill high proof alcohol the temperature should be as low as possible without going below the point where all the alcohol will condense. Exact temperature depends on altitude and a variety of other factors. An automatic valve will control this accurately.

CONDENSER SECTION

This is simply a coil of tubing with cool water passing through it to cool the vapors and condense everything that has gotten through the stripper section. The bottom of this section funnels the vapors into the cooling area around the coils and forms a trough for the condensate to flow out of the still.

COOLING WATER PLUMBING

This can be hooked up almost any way and still work. However, because this is one of the points at which energy is put into the still we should see what we can do to make it even more energy efficient. There isn't any sense in cooling the alcohol below 100° F. It becomes fairly stable at this point. Therefore the flows of water in the condenser coil needs to be no greater than to cool the liquid to 100°F. This flow will be determined by the amount of vapor to be condensed, the amount of water in those vapors and the temperature of the water going in.

The flow of water through the stripper will be determined by the proof we want. The more vapor going through the still the more water needs to flow through the stripper to keep the temperature of this section constant. The water going into the stripper should be the water coming out of the condenser. This water is already warmed somewhat and is less likely to overstrip or condense the alcohol vapor. The water should enter at the top and come out the bottom of the stripper coils.

The water coming out of the stripper will be at a very high temperature. It can be used as a heat source for the distilling room, another room or part of the alcohol production process. It could also be used as domestic hot water.

MATERIALS AND SPECIFICATIONS.

Study the material's list and drawings for each section as you read the following so you will be able to purchase the correct parts and put them together so the still will function properly.

Parts Sheet - Each part is given an identification number. Most of the dimensions are given on this sheet. In cases where more than one part of the same material is used in one section the total for that section is given on the parts sheet and the dimensions of each piece are given on the scale drawing of the section. The "Total" column on the parts sheet gives the amount of the item you will need for the entire column. This can be used as your shopping list. However be sure to look over the sheet showing the miscellaneous parts to be sure that you get enough sheet copper to be able to cut all four plates and the funnels out. You should buy 8 feet of 3" pipe in order to allow some extra material for uneven cuts or mistakes.

Trap - The runoff line sets the depth of the pool inside both the trap and the doubler section. It should be 2" from the top of the trap and may face whatever direction is most convenient for your still. You may connect any sort of line to piece A-8 as long as it can withstand temperatures around the boiling point of water. The material you choose will determine what sort of fitting you attach to A-8 or whether you will simply use a hose clamp.

The coupling between the trap and the doubler shows a T (A-10). This allows you to drain the pools. If you do not wish to be able to drain the pools you may omit this piece and make a straight coupling with a piece of ½" pipe.

Silver solder the A-8's into the 3" pipe and soft solder the other parts. Part A-2 should not be soldered. In fact a very small hole in the top will vent the section and prevent any siphoning from occurring.

Doubler - The vapor line can enter from any direction, to be determined by where your boiler will set in relation to your still. The vapor line must enter close to the bottom of the section, however, to allow the vapor to gain as much benefit as possible from passing through the pool.

Silver solder B-3, B-4 and B-8. Soft solder the other parts. This allows the other parts to be put on or removed without something else falling off. The 3" copper pipe will conduct enough heat from where you are soldering to parts already soldered to melt anything but silver solder.

Stripper - This is a very simple section to construct. Simply silver solder C-3 onto the section. The S-1 plates and S-3 stripper materials are simply laid into place before joining the two sections together with soft solder. For an even higher capacity, more effective still a coil of copper tubing can be placed in this section and the water from the stripper coil above it run through it for additional stripping action. That coil can be made in the same way as the stripper coil and condenser coil, although it will need to be longer. It should be in place before the stripper material is placed in the still.

Stripper Coil - The coils in this section should be formed by wrapping the tubing around a 1 1/4" iron pipe. This diameter places the coils more centrally in the section so that the condensate will drip back onto the plate for even distribution.

Make the S-12 fittings, making sure that the 1/4" tubing will fit snugly into the hole after the end of the S-12 is pinched shut and silver soldered. Notch out the D-3 so as to clear the S-12. Silver solder the D-3 and S-12 at this end then connect and silver solder the coil at this end. Next finish the other end, silver soldering the S-12 and coil.

Notice that the D-13 is an elbow and the D-15 is a pipe adapter. The control valve will be connected to D-15 and the coolant flow from the condenser will be connected to D-13. Both connections can be placed in any direction desired. You may want to offset D-13 to allow for straight connection with F-13, avoiding the water hookup line, F-12. Otherwise you can bend the tube around this connection.

Control - Silver solder E-3 onto E-1 then bore and fit the other parts. Silver solder E-12 into E3 carefully so as not to cause any problems in getting the funnel into place. Silver solder E-16 and E-7. Again, these parts may face any direction. This area can get rather congested when all the lines are connected so plan the placements carefully.

Soft solder the other parts together. The funnel, like the plates, is simply set into place during assembly.

Condenser - This is very similar to the stripper coil section except that it is at the top of the column. It is longer and both S-12's are placed on the same side. The coils should be wrapped around a 1 ½ " pipe. This leaves more room to clear the top of the funnel. Watch the length of the lower S-12 and put the section in carefully so it will not hit the funnel.

Soft solder the F-12 on. If the coolant water should go off the vapor should go out the vapor line This thin cap should be the first thing to blow if something should go wrong with the column.

Be sure to make a notch (see note 4) for the alcohol line.

Miscellaneous Parts - Plates S-1 don't have to be very thick as they are quite strong when installed. Don't worry about making nice round holes. Oblong holes are better because the marbles cannot settle into them and seal them off.

The funnel is best made by slowly bending until it will sit in the control section. The hole for the 3/4" pipe will be too small. Working this open over something will swell it to make a nice solder point for soldering the base to the pipe. This connection should be silver soldered.

Completing the Column - When all the sections are finished, stack them together in order making sure the finished column will be straight and soft solder round each part. Soft solder A-9 to B-8 and you are ready to test your column.

Vent
Condenser
Coolent In
Thermometer
Alcohol
Control Probe
Control
Stripper
Coil
Plate #2
Coolent Out
Control Valve
Stripper
Plate #1
Trap
Runoff
Vapor In
Doubler
and
Trap
Pools
Drain

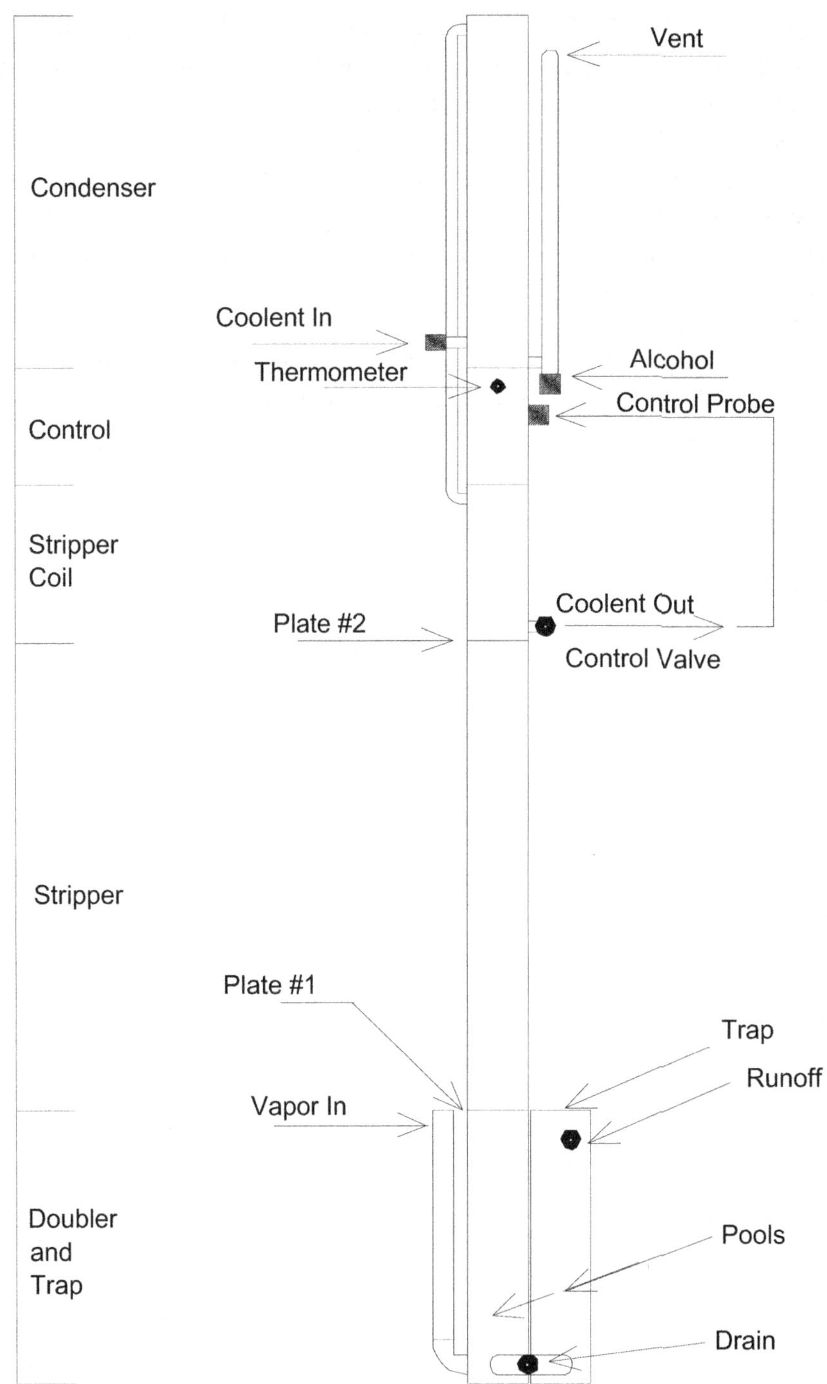

Scale 1/8" = 1"

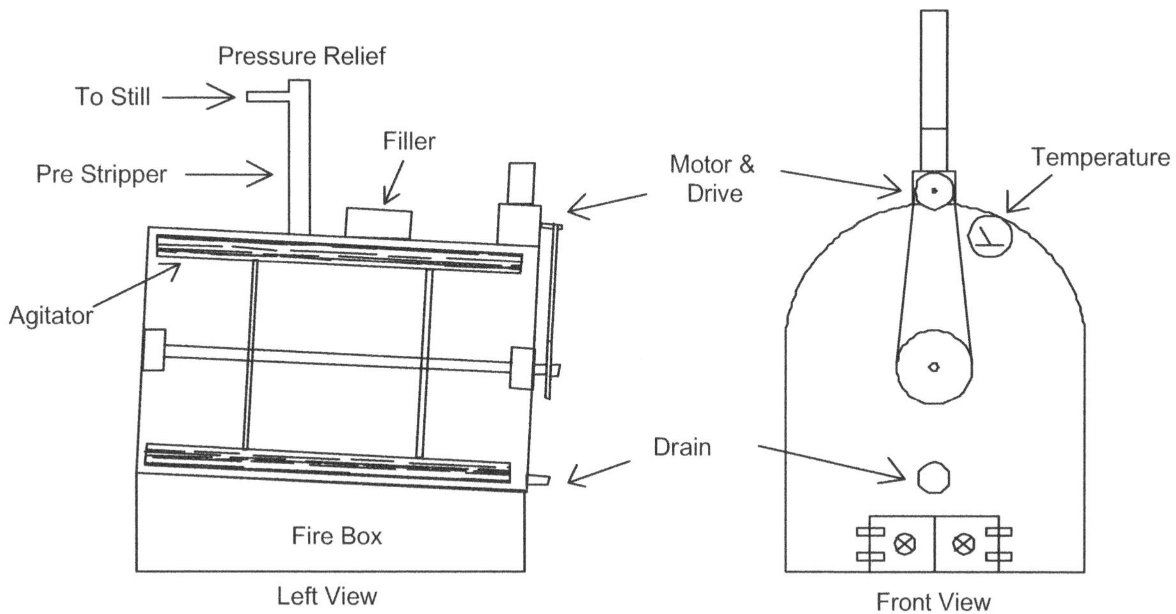

Cooker-Fermenter-Boiler

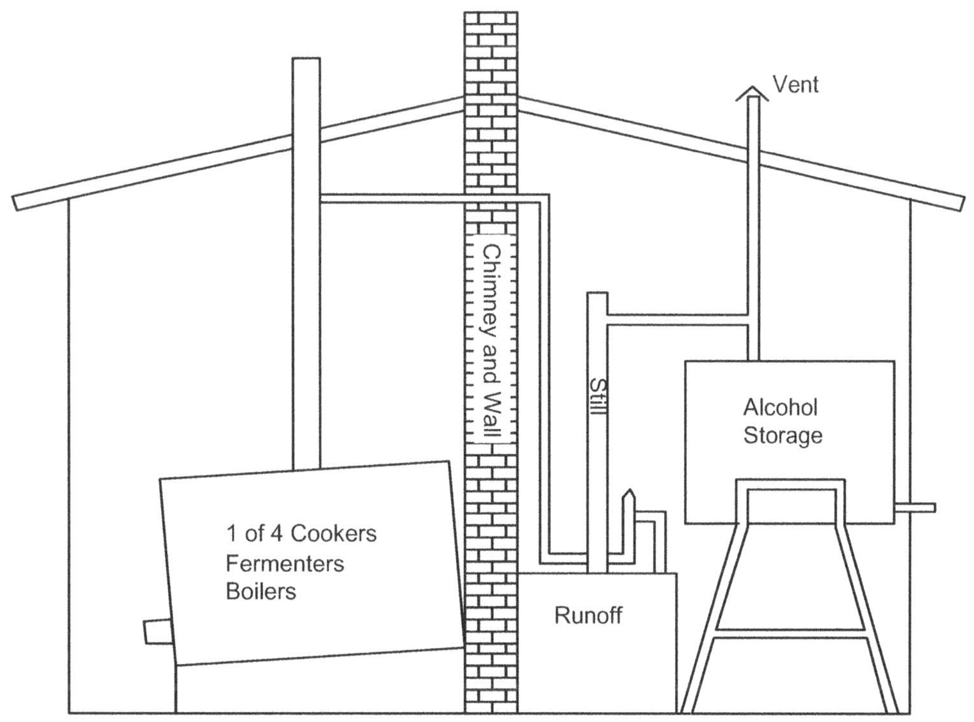

Typical Farm Fuel Plant

#	Trap	Doubler	Stripper	Stripper coil	Control	Condenser	Total	Description
1	14"	14"	24"	10"	6"	18"	7'2"	3" DMV (pipe)
2	2	1				1	4	3" DMV (test cap)
3		1	1	1	1		1	3"DMV (repair coupling
4		14"			1 1/2"		151/2"	3/4" copper pipe
5		2					2	3/4 elbow
6		1					1	3/4s x 3/4mp adapter
7					1		1	3/4s x ½ fp adapter
8	2"	1"					3"	½" copper pipe
9	2						2	½" Street Elbow
10	1						1	½" T
11					1		1	3/8" T
12				3"	16"	3"	22"	3/8" copper pipe
13				1		1	2	3/8" elbow
14						1	1	3/8"s x ½"mp adapter
15				1	1		2	3/8"s x 3/8" mp adapter
16				19-21 turns	3 ½"	34-36 turns	50'	1/4" refrigeration copper
S-1	1	2	1				4	3 1/8d copper plate
S-2					1		1	3 3/8d copper funnel
S- 2A					1		1	3/4" pipe x 1 1/2"
S-3			9 lb.					5/8" glass marbles
S-12				2		2	6"	3/8" copper x 1 1/2

DOUBLER - B

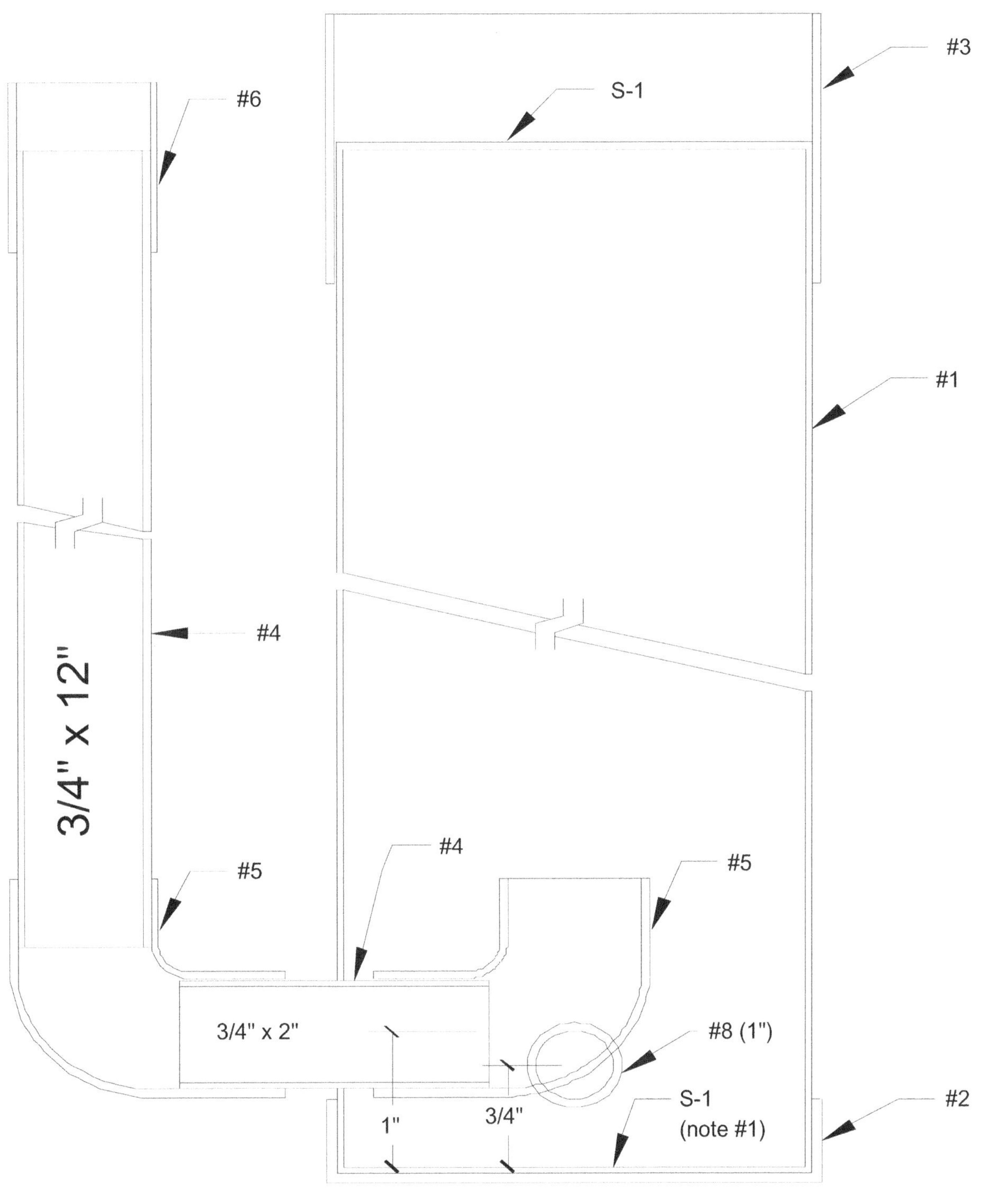

STRIPPER - C

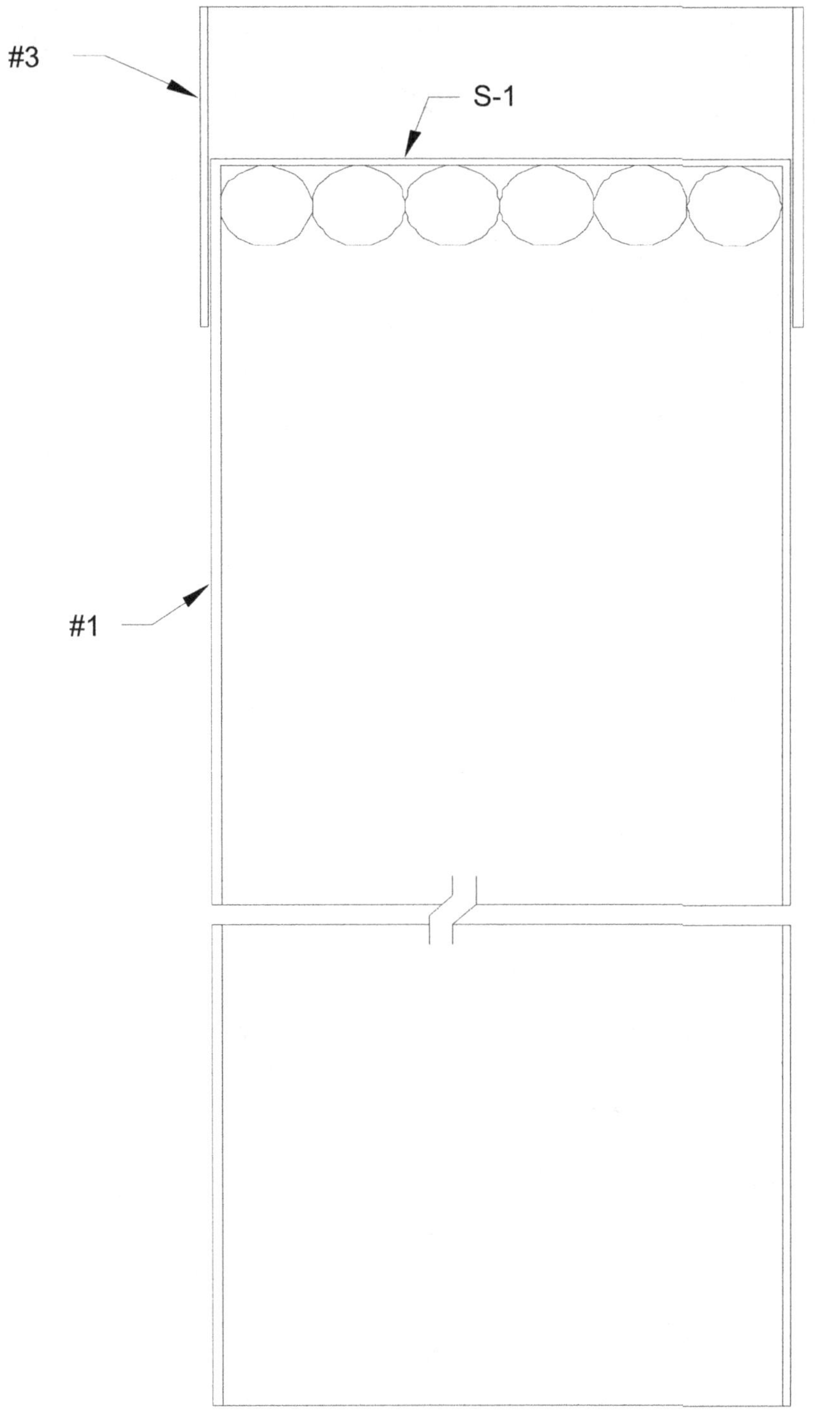

STRIPPER COIL - D

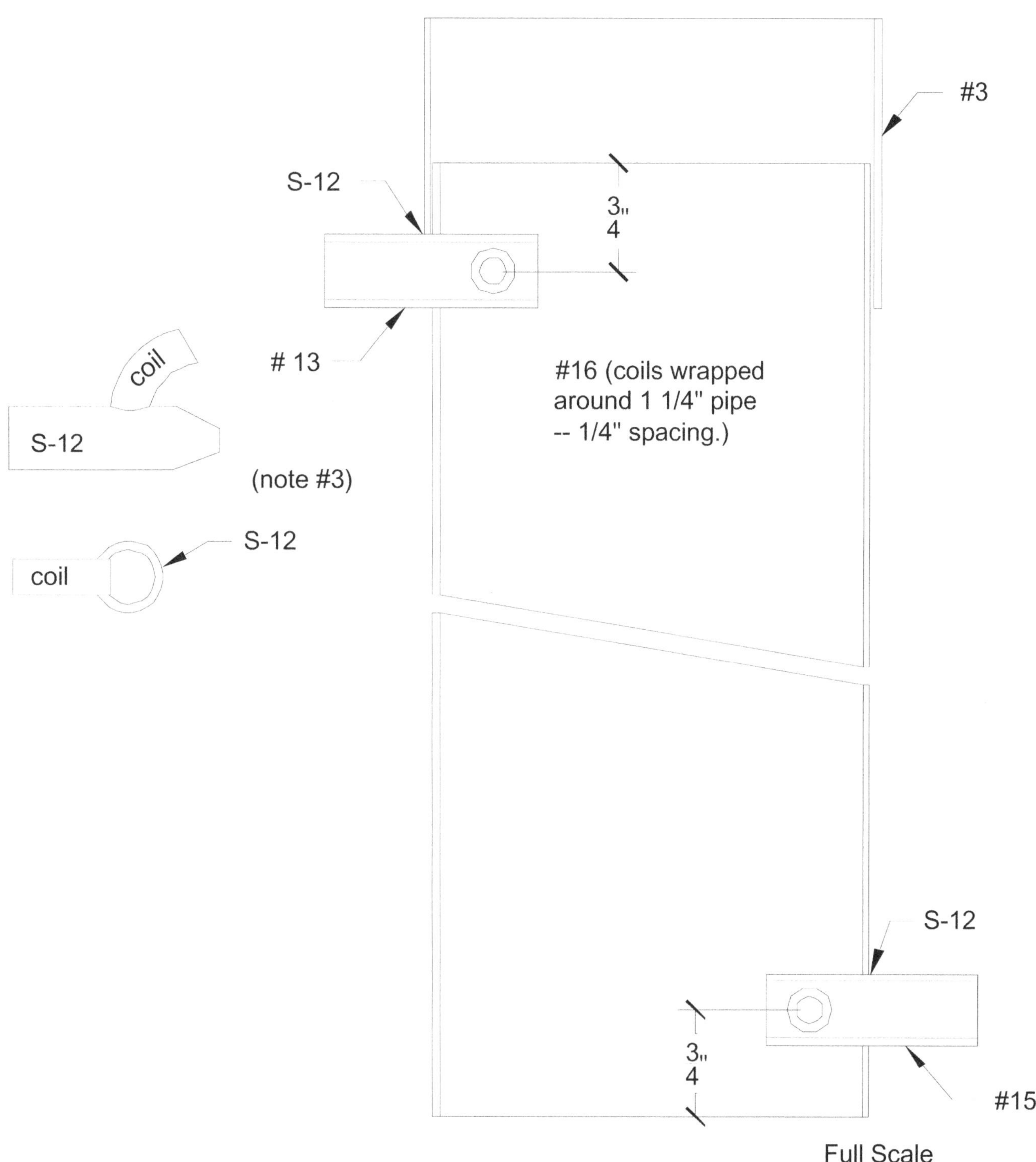

CONTROL - E

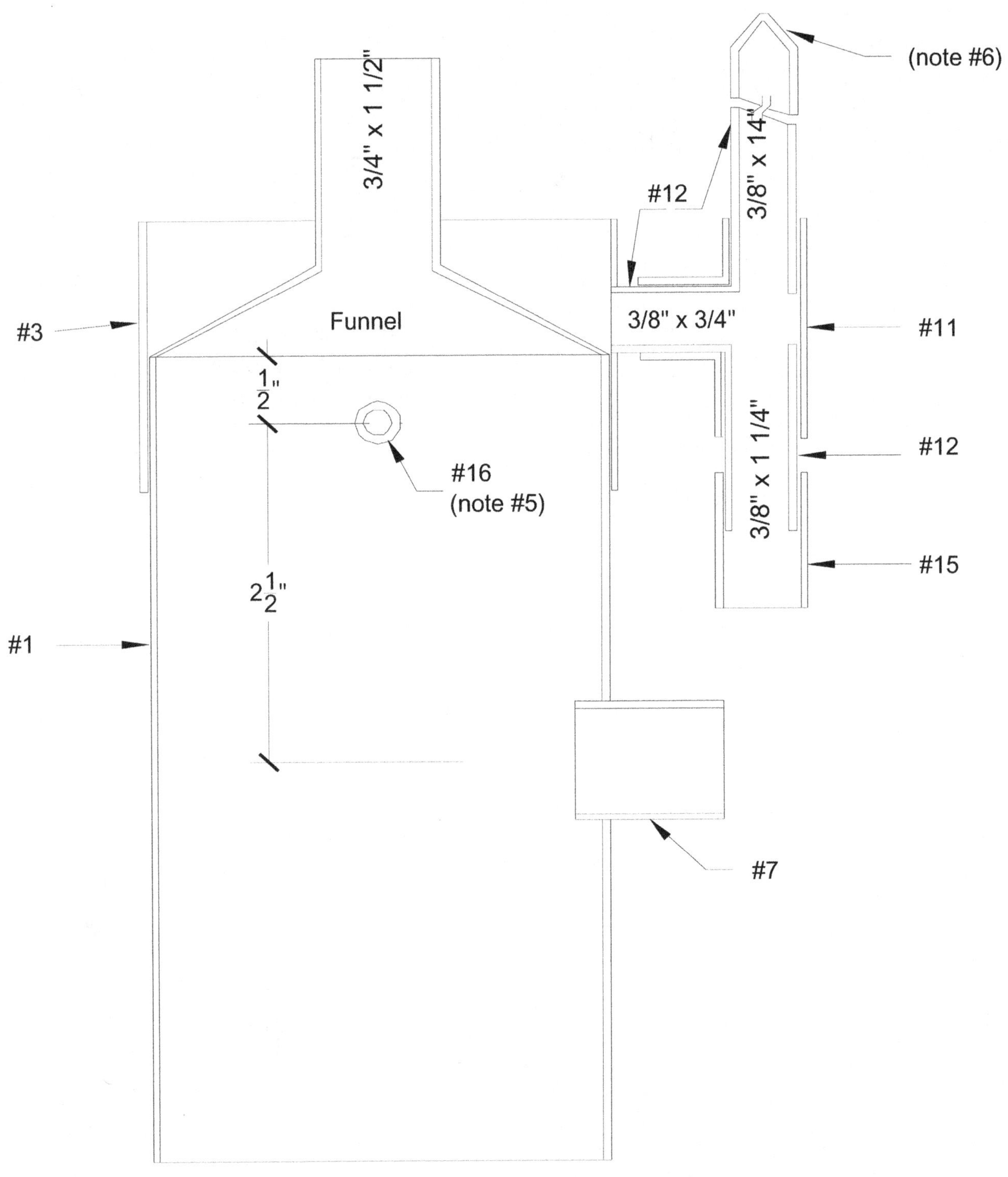

116

CONDENSER - F

3/4"
S-12
#13
#1
#2
#16 (coils wrapped
around 1 1/2" pipe
-- 1/4" spacing.)
#14
S-12
1 1/2"
Note #4
Full Scale

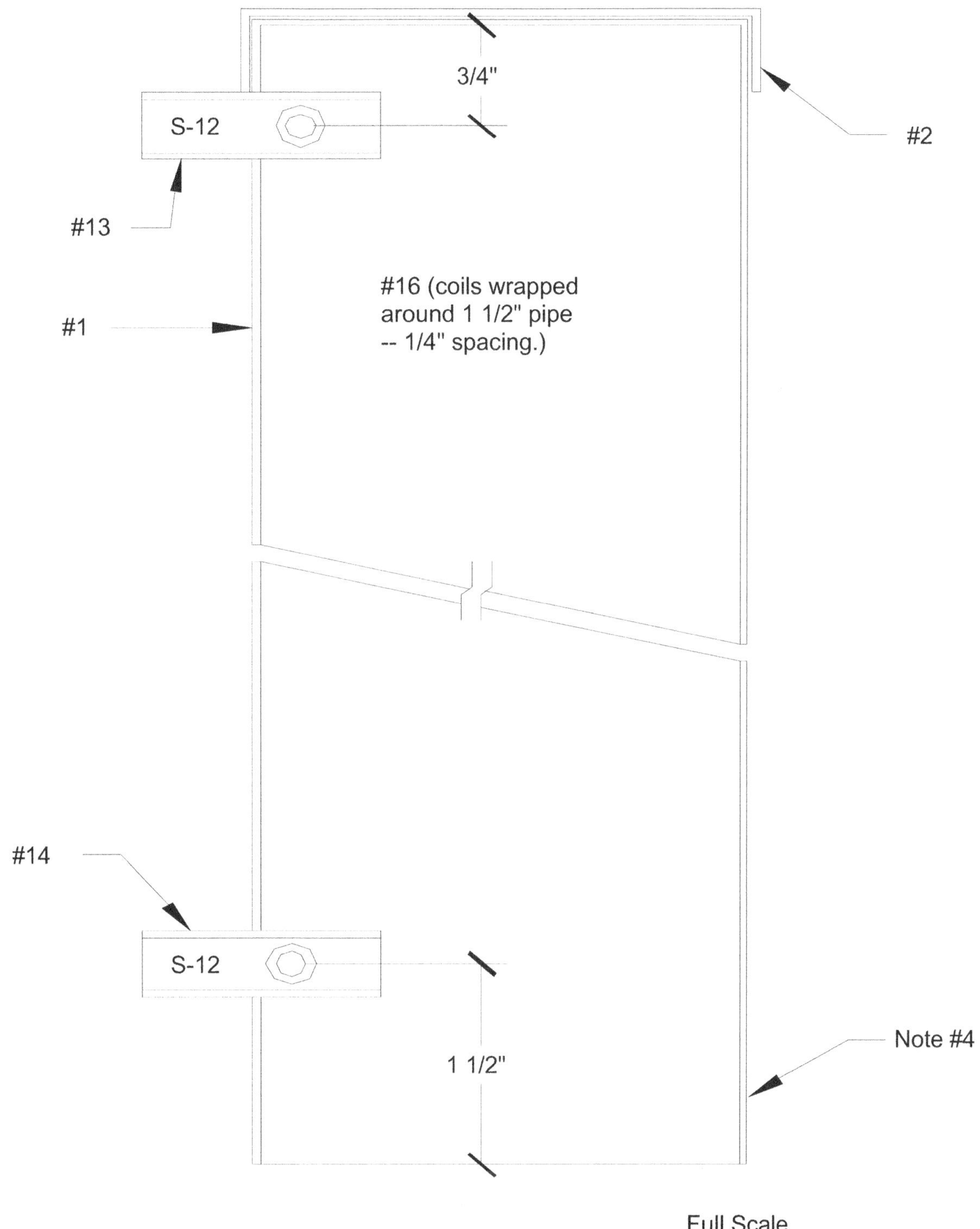

MISC. PARTS

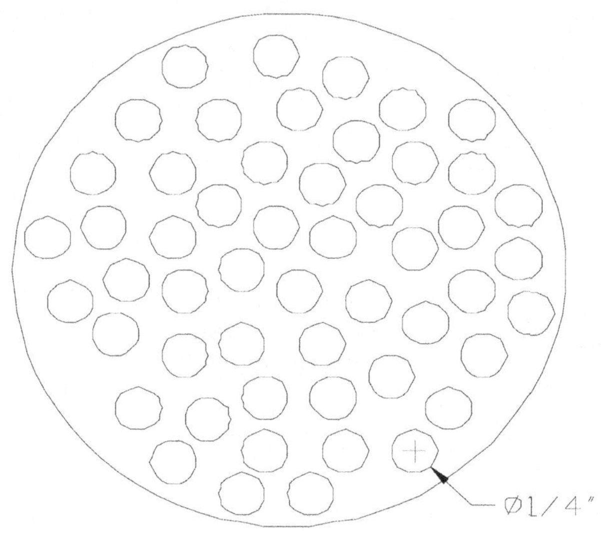

S-1

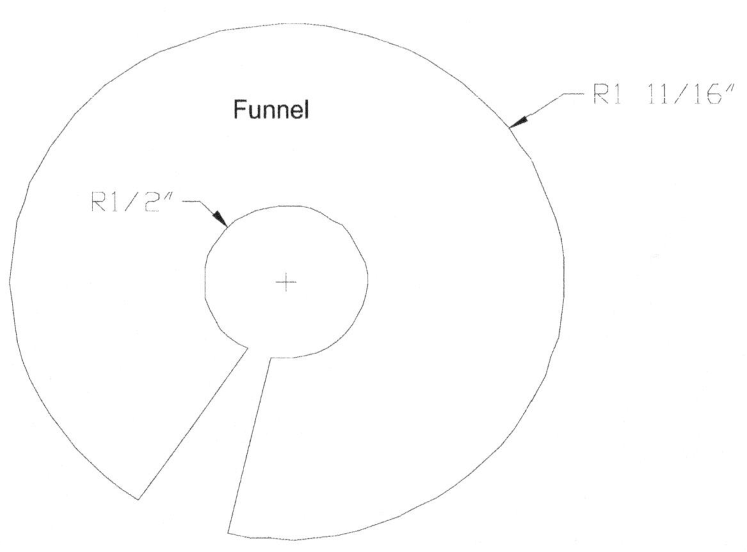

Full Scale

NOTES

#1   This is an S-1 plate without the holes. The thin test caps need some support on the bottom. Simply cut out an S-1 and place it in the cap. The S-1 plates can be made from split 3" pipe if flat copper is not available.

#2   This is a T in the coupling line for a drain that will empty both pools. This may be closed by either a valve or a plug. Normally the pools will not need to be drained and this part may be omitted if you feel it is not necessary. If there is any possibility that the still will freeze or that you will want to move it be sure to include the drain.

#3   This diagram demonstrates an easy way to get the outlet of the cooling coils through the 3" pipe. Simply bore a 1/4" hole in the S-12, make sure the 1/4" coil fits. Pinch the end of the S-12 shut as tightly as possible and silver solder. Make sure everything fits then silver solder the S-12 into the 3" pipe. Solder the coil in last and water check. <u>CAUTION</u>: Be sure the end of the coil is reamed clean and does not extend into the S-12 any further than necessary. Either situation could cause added back pressure on the coolant flow.

#5   This would normally be 1 1/4" tube pinched off and soldered like part S-12. However, should you choose a probe thermometer with threads it will require a fitting like E-7 of the proper size. If the probe diameter is larger than 1/8" you will need to use a larger tube. Make sure that the end of the tube is sealed well so there will be no leaks if the thermometer is removed.

#6   This T and vent line were not used on earlier columns and this lack caused no problems. However, if the alcohol line should happen to get covered by the alcohol filling the storage container a vacuum will be formed that will draw both the water from the trap and some of the alcohol from the container into the column. This will stop shortly then alcohol and water will come rushing back out. The still will continue to function in alternate cycles of drawing in and forcing out. Sometimes there is a loud popping sound as this begins. The still will continue to distill alcohol but the sound can be really frightening.

THE STILL IN OPERATION

With a thermometer in the liquid area of the boiler we can watch the temperature rise in the boiler. At some temperature above 173° F. the pipe between the still and boiler will begin to heat up. The exact temperature at which this happens will depend on the concentration of alcohol in the brew. A low boiling point indicates a high concentration of alcohol. You can follow the vapors be feeling the pipe and still as they warm up.

REMEMBER - Distillation starts at the boiler and the vapors are being cleaned as they travel.

Soon the temperature in the control section will start to rise. When this occurs start a cooling flow of water. If you are controlling the flow manually let the temperature rise to around 190°F. and hold it there to start with. Take a proof reading every 30 seconds until the proof starts to level off. Once it has leveled off you can start more water flowing through the coils to get the proof you desire. The less vapor flowing through the still the smaller the change and the slower the reaction will be.

Be sure you allow time for the proof to level off after each adjustment. Record the proof and temperature with each adjustment so you can go to that temperature for that proof.

An automated valve that operates at temperatures between 173°F. and 212°F. will make it much easier to control the proof. If you have an automatic valve turn it to the lowest temperature setting. When the control section temperature stabilizes, slowly turn the valve to a higher setting until the alcohol starts coming out. Let the still stabilize here and check the proof. Continue making adjustments until you reach the proof you want. Once you are familiar with the valve you can set it quickly to get whatever proof you like and you will consistently get that proof. Use the chart on page 59 to determine the actual proof of the alcohol if you measure it while it is still warm.

Don't be surprised if alcohol starts coming out of the still before you turn the water on. The vapors will condense on the condenser coils and the side of the condenser as they warm the still.

REMEMBER - The control temperature is the result of vapor flow. If you don't have any vapor flow from the boiler the control temperature will not be very high. The more alcohol in the beer the sooner the vapor begins to flow. If there is no alcohol n the beer you will get nothing but water vapor and nothing will come out the alcohol outlet, provided the controls are set properly.

Soon after you start distilling there will be some flow out the runoff outlet. This will mean that the pool is full and the still is nearly stabilized. Once the pool is full and you have adjusted the temperature to the proof you want the alcohol flow will increase. The up and down flow is established and the pool is flashing the alcohol that reaches it back up to the stripper.

REMEMBER - The brew is losing alcohol. The smaller the batch or the more heat that is being applied the faster this happens.

As the alcohol percentage falls the temperature in the boiler will begin to rise. This is because the mixture has more and more water in it as the water vapor condenses and falls back into the tank and less alcohol as the alcohol vapor flows through the still. The rise in temperature will be seen in the still, too, so little by little more water will need to be sent through the stripper to hold that temperature, and therefore the proof of the alcohol, steady. This is where it is important to have an automatic valve to hold the proof steady. Watch the boiler temperature closely and record the temperature and the length of time it takes for this rise to occur.

The operation of this still is such that the heaviest flow of alcohol will be at the beginning and it will slowly taper off until the alcohol stops altogether. Check the proof as it gets near the end. If it begins to fall increase the stripper water. You will notice more and more flow out the runoff. When the alcohol all but stops, check the boiler temperature. This is the temperature at which the still will have most of the alcohol out of the brew. Turn off the heat to the boiler at this time.

The liquid in the boiler contains the acid you added earlier to lower the pH of your brew. This water may be cooled, the solids settled out and fed to hogs, poultry, or fish and the remaining liquid put in the fermentation vat as the second addition of water to cool the brew before adding the second enzyme.

RUNOFF

There will always be some alcohol in the runoff. Although the quantity may be slight it is a good idea to rerun the runoff. It is much cleaner than the original brew and will clean out your boiler and plumbing on the way to the column.

CORRECTIONS AND ADJUSTMENTS

The first thing that must be taken care of if it exists when you first get your still in operation is any liquid or vapor leak in the still. Vapor loss means loss of alcohol. This can be checked before operation by sealing all but one opening in the still. Blow air into the still. Any air leak would be a potential vapor leak. Seal these up.

If little or no alcohol came out and the temperature in the boiler slowly rose to the boiling point of water (212° F.) the problem is most likely not with the hardware but with the fermentation. Be sure to check the runoff. There may not have been enough heat to get the alcohol to the top of the still. With too small a heat supply this could happen. The still is designed the size it is so that you can use heat sources that are normally found on the farm to power it. It can be run at minimum capacity with a burner from a gas or electric stove. This will produce 6,000 to 10,000 BTU/hour of heat.

Failure of alcohol to reach the condenser section could also result from too great a heat loss. If you are operating the boiler outside in winter just a little wind could cause too great a heat loss and all the alcohol could be in the runoff at fairly high proof or could have run back into the boiler. The boiler and still should both be inside a building.

TOO HIGH AN OPERATING TEMPERATURE

If the control temperature is much above 173° F. when the proof of the product has peaked out (should be above 190 proof) you should make some corrections. This is most likely caused by too much back pressure in the boiler. Although the pool controls the back pressure it should not be changed unless your pool is very deep. Within reason, the deeper the pool the lower the proof of the runoff. Too great a distance between the boiler and the still with too small a pipe is the most likely cause of the problem. The easiest way to maintain the low pressure required (5 lb.) and at the same time provide a blowoff valve for your boiler is to come out of the boiler with a large pipe. This pipe should be 1 ½ to 3 inches depending on the size of your boiler. Run this pipe up and out of the top of the building. Put a metal plate on top of the pipe with a rubber ring under it. Weight the metal plate until it weighs 5 pounds. A tractor exhaust rain cap with a rubber seal will work nicely. This provides a pressure release at a safe distance from workers and visitors who seem to inevitably gather around a still and also provides some stripping action. The line to the still can be tapped off this pipe and will pick up vapors with a higher concentration of alcohol higher up in the still.

Made in the USA
Monee, IL
07 July 2026

56550066R00077